A TEXT BOOK OF
QUANTUM MECHANICS

A TEXT BOOK OF QUANTUM MECHANICS

By
M.G. Arora

ANMOL PUBLICATIONS PVT. LTD.
NEW DELHI - 110 002 (INDIA)

ANMOL PUBLICATIONS PVT. LTD.
4374/4B, Ansari Road, Daryaganj
New Delhi - 110 002
www.anmolbooks.com

A Text Book of Quantum Mechanics

First Edition, 1997
Reprint, 2002

ISBN 81-7488-458-0

PRINTED IN INDIA

Published by J.L. Kumar for Anmol Publications Pvt. Ltd., New Delhi - 110 002 and Printed at Mehra Offset Press, Delhi.

Contents

Preface *vi*

1. Fundamental Concepts of Quantum Mechanics 1

2. Waves and Quanta 45

3. Wave Equations and Solutions for Some Simple Systems with Constant Potential Energy 63

4. Wave Mechanics of Some Simple Systems where the Potential Energy is not Constant 89

5. Various Solutions of Schrodinger's Equation for Hydrogen Like Atoms 108

6. Quantum Chemistry and Valency 149

7. Hückel Molecular Orbital Theory 188

Preface

The present book on quantum mechanics is intended to serve undergraduate and post-graduate students. The book has been divided into two parts. The Part I deals with theoretical concepts in quantum mechanics while Part II deals with problems based on quantum mechanics. Most of the problems have been taken from various university examinations. S I units have been used. However, at some places old units have been used so that the students may become familiar with old as well as new units. Throughout this book, it is assumed that the student understands the fundamental concepts in physical chemistry. Each concept/problem covered in this book can do full justification for most of the students.

I am thankful to my several immediate colleagues and friends for making useful suggestions during the preparation of manuscript of this book.

The author expresses his sincere thanks to Dr. A. Kumar whose constant inspiration has initiated the author to bring out this book.

Suggestions for further improvement towards this book shall be gratefully acknowledged and incorporated in the next edition.

—Author

1

Fundamental Concepts of Quantum Mechanics

Introduction

Quantum mechanics often seems very abstract in its formulation, but it is not merely a system of mathematics. It is needed to describe the results of experimental work. It has been devised to describe the results of observations and experiments, made in the study of atomic physics where it is found that a straightforward application of ordinary means of description leads to inconsistencies and contradictions. For this reason it is desirable to begin the study of quantum mechanics with a compilation and analysis of some of the principal types of experiments that have led to the development of the theory, and it is helpful at all times to keep in mind the physical implications of the various features of the mathematical formulation.

Quantum mechanics is a theory, of which the classical mechanics is a limiting case, that suffices to correlate in a unified scheme the variety of experimental results described in this chapter, as well as many others. The remainder of this book will be devoted to the presentation and application of a set of concepts and principles in terms of which all of these experiments can be described. No attempt will be made to derive the principles from the observations. The principles are obtained by induction and assumption, not by deduction. No claim is made that the quantum mechanics to be presented is the only theory that could be devised. In fact, several different, but equivalent, forms of the theory will be presented. Each form has its advantages, and each can be used where its application makes calculation simplest.

Although the quantum mechanics is adequate for the treatment of a wide range of phenomena that are beyond the range of classical mechanics, there are still limits to its validity and cases in which it is not completely satisfactory. Some of these will be pointed out from time to time as they arise. Nevertheless, the emphasis will be on the successes of the theory, since it has made possible for the first time an adequate electron theory of matter and has made possible an extensive, if not entirely adequate, understanding of atomic nuclei.

The mathematical language of quantum mechanics is geometrical in nature. This statement may strike the beginning student as curious if he has ever peeped inside the cover of a textbook on the subject and been frightened by the complexity of the analytical formulas that disfigure almost every page. Essential to a sound grasp of the subject, however, is an ability to generalize one's intuitive perceptions of the geometry of two and three dimensions to spaces of a higher number of dimensions. The analytical tool necessary to do this is vector algebra.

Vectors

A *scalar* is an ordinary number, real or complex. A *vector* **x** is a set of scalars written out in some order, either as a row or as a column:

$$\mathbf{x} = (x_1, x_2, \ldots\ldots, x_n)$$

The number n of *components* (entries) of the vector is said to be its *order* or *dimensionality*.

A geometric picture of a vector is conveniently obtained for $n = 2$ or $n = 3$ by interpreting the components of the vector as the coordinates of a point in n dimensional space. Then the directed line segment drawn from the origin of a Cartesian coordinate system to the point in question is a geometric realization of the vector (Figure 1.1).

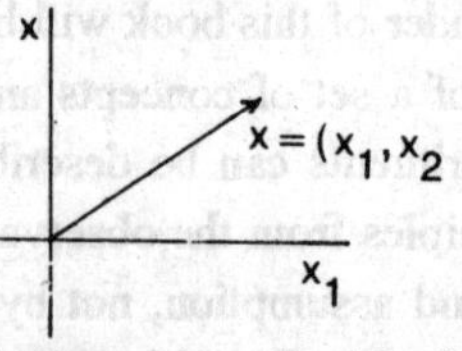

Fig. 1.1. *Geometric realization of a vector.*

Algebraic Properties of Vectors

Two vectors are said to be equal only if they are of the same order and have all their components separately equal. Thus **x** = **y** implies $x_1 = y_1$; $x_2 = y_2$;..... ; $x_n = y_n$, so that a single vector equality is equivalent to n scalar equalities.

Vector addition is defined by the equation

$$\mathbf{x} + \mathbf{y} = (x_1, x_2, \ldots\ldots, x_n) + (y_1, y_2, \ldots\ldots, y_n)$$
$$= (x_1 + y_1, x_2 + y_2, \ldots\ldots, x_n + y_n)$$

It follows that vectors can be added only if they are of the same order and that vector addition is commutative: $\mathbf{x} + \mathbf{y} = \mathbf{y} + \mathbf{x}$. In two dimensions, vector addition is identical with the well known parallelogram rule of Galileo (Figure 1.2).

The *null* vector has all of its components identically zero : $\mathbf{0} = (0, 0, \ldots 0)$, and we have the theorem $\mathbf{x} + \mathbf{0} = \mathbf{x}$.

The product of a vector by *a* scalar a is defined by

$$a\mathbf{x} = (ax_1, ax_2, \ldots\ldots, ax_n) = \mathbf{x}a$$

Illustrative of these definitions are the examples

(1) $(-1, 0, 3) + (1, -1, -4) = (0, -1, -1)$

(2) $2(0, -2, 2) = (0, -4, 4)$

(3) $4(1, -1, 0) - 2(1, -2, \frac{1}{2}) = (2, 0, -1)$

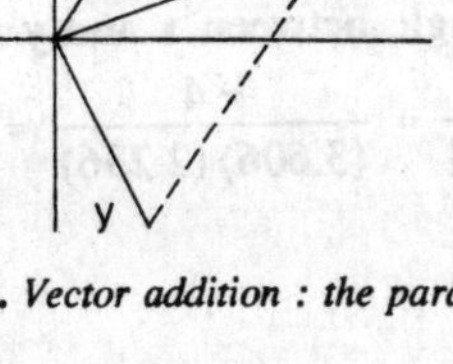

Fig. 1.2. *Vector addition : the parallelogram rule.*

The Scalar Product

The *scalar product* of two vectors is not a vector but a scalar and is defined only for two vectors of the same order:

$$\mathbf{x} \cdot \mathbf{y} = x_1y_1 + x_2y_2 + \ldots + x_ny_n = \sum_1^n x_i y_i$$

Corresponding components of the two vector are multiplied and the products summed. Note that $\mathbf{x} \cdot \mathbf{y} = \mathbf{y} \cdot \mathbf{x}$.

The scalar product of a vector by itself is the sum of the squares of its components. The positive square root of this sum is called the *norm* of the vector:

$$|x| = (\mathbf{x}\cdot\mathbf{x})^{1/2} = (x_1{}^2 + x_2{}^2 + \ldots + x_n{}^2)^{1/2} = \left[\sum_1^n x_j{}^2\right]^{1/2}$$

When interpreted geometrically in two and three dimensional space, the norm of a vector is by the Pythagorean theorem its length. It can also be demonstrated that the scalar product of two vectors in two or three dimensions is

$$\mathbf{x}\cdot\mathbf{y} = |\mathbf{x}|\,|\mathbf{y}|\cos\theta$$

where $0 \le \theta \le \pi$ is the angle between the two vectors. A vector of unit norm $|x| = 1$ is said to be a *normalized* or *unit* vector.

If $\mathbf{x}\cdot\mathbf{y} = 0$, the vector **x** and **y** are said to be *orthogonal.* From the formula $\mathbf{x}\cdot\mathbf{y} = |x|\,|y|\cos\theta$, we observe that this happens in two and three dimensions when $\theta = \pi/2$ radian $= 90°$, that is, when the vectors are perpendicular. Note that formally $\mathbf{x}\cdot\mathbf{0} = 0$ for every **x**, so that the null vector is orthogonal to every other vector.

To illustrate these operations, consider in Figure 1.2

$$\mathbf{x} = (2, 3); \qquad \mathbf{y} = (1, -2)$$

The scalar product is $\mathbf{x}\cdot\mathbf{y} = (2)(1) + (3)(-2) = -4$. The norms are

$$|x| = (2^2 + 3^2)^{1/2} = 3.606$$

$$|y| = [1^2 + (-2)^2]^{1/2} = 2.236$$

The cosine of the angle between **x** and **y** is therefore

$$\cos\theta = \frac{\mathbf{x}\cdot\mathbf{y}}{|\mathbf{x}|\,|\mathbf{y}|} = \frac{-4}{(3.606)(2.236)} = -0.496$$

or $\theta = 119°45'$

Matrices

In this book a *matrix* will always be a square array of n^2 scalars

$$A = \begin{pmatrix} a_{11} & a_{12} & \cdots & a_{1n} \\ a_{21} & a_{22} & \cdots & a_{2n} \\ \cdot & \cdot & & \cdot \\ \cdot & \cdot & & \cdot \\ \cdot & \cdot & & \cdot \\ a_{n1} & a_{n2} & \cdots & a_{nn} \end{pmatrix}$$

arranged in some order, where the number n of rows or columns is the *order* or *dimensionality* of the matrix. Two matrices are said to be equal only if they are of the same order and have corresponding *components* (entries) identical. Thus $A = B$ implies $a_{ij} = b_{ij}$ so that a single matrix equality is equivalent to n^2 scalar equalities.

Matrices exist in which the number of rows is not equal to the number of columns, but such rectangular arrays will not be needed in this book.

Matrix Algebra

Matrices of the same order may be added to yield another matrix under the rule $A + B = C$ implies $c_{ij} = a_{ij} + b_{ij}$, that is, corresponding components of A and B are added to yield C. The *null* matrix θ has all of its components equal to zero, and for this matrix $A + \theta = A$.

For vectors we defined a scalar product such that the result of the multiplication is not a vector but a scalar. For matrices, however, we define a matrix product such that two matrices of the same order may be multiplied together to yield a third matrix. If $AB = C$, then the arithmetic rule whereby the components of C are calculated from the components of A and B is

$$c_{ij} = \sum_{k=1}^{n} a_{ik} b_{kj}$$

Note that c_{ij}, which is the component of C located at the intersection of the ith row with the jth column, is calculated from the components of the ith row of A and the components of the jth column of B (Figure 1.3).

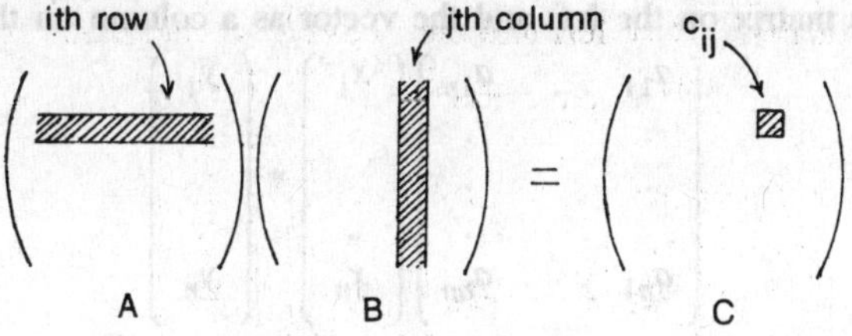

Fig. 1.3. *Matrix multiplication.*

The student unfamiliar with the arithmetic details of matrix multiplication will do well to work out a few examples. Thus if

$$A = \begin{pmatrix} -1 & 0 & 1 \\ 2 & 3 & 0 \\ 0 & -2 & 1 \end{pmatrix}$$

$$B = \begin{pmatrix} 7 & 0 & 2 \\ -6 & 3 & 1 \\ 4 & 0 & -3 \end{pmatrix}$$

then c_{11} is calculated from the first row of A and the first column of B:

$$c_{11} = (-1)(7) + (0)(-6) + (1)(4) = -3$$

Similarly

$$c_{12} = (-1)(0) + (0)(3) + (1)(0) = 0$$

$$c_{13} = (-1)(2) + (0)(1) + (1)(-3) = -5$$

The student should continue this process to find

$$C = \begin{pmatrix} -3 & 0 & -5 \\ -4 & 9 & 7 \\ 16 & -6 & -5 \end{pmatrix}$$

It cannot escape him that the numerical procedure in calculating a matrix product is identical in vector algebra with that of forming n^2 scalar products.

Multiplication of a Matrix Times a Vector

The product of a matrix times a vector is another vector:

$$A\mathbf{x} = \mathbf{y}$$

where $y_i = \sum_{j=1}^{n} a_{ij}x_j$. It is conventional in carrying out this operation to write the matrix on the left and the vector as a column on the right:

$$\begin{pmatrix} a_{11} & \cdots & a_{1n} \\ \cdot & & \cdot \\ \cdot & & \cdot \\ \cdot & & \cdot \\ a_{n1} & \cdots & a_{nn} \end{pmatrix} \begin{pmatrix} x_1 \\ \cdot \\ \cdot \\ \cdot \\ x_n \end{pmatrix} = \begin{pmatrix} y_1 \\ \cdot \\ \cdot \\ \cdot \\ y_n \end{pmatrix}$$

Then the arithmetic follows the same row-column pattern as matrix multiplication in that y_i is calculated from the elements of the ith row of A and those of the single column of $\mathbf{x}$. Note that a matrix enters into mathematics as an operator which sends a vector into another vector, changing its direction as well as its length. We shall see that the mechanical state of a system in quantum mechanics is described by a vector. Only those vectors are acceptable as state vectors that transform in a definite way when operated on by certain matrices.

An important theorem concerning the multiplication of a matrix times a vector is that the operation is distributive:

$$A(\mathbf{x} + \mathbf{y}) = A\mathbf{x} + A\mathbf{y}$$

Linear Transformations and Matrices

The totality of all vectors of order n is known as an n dimensional vector space. Because the result of operating on a vector with a matrix is another vector, the result of operating on a vector space with a matrix is to map the space onto itself. Such a mapping is known as a *linear transformation.* Let us consider some examples.

Suppose that

$$D = \begin{pmatrix} 2 & 0 \\ 0 & 2 \end{pmatrix}$$

and consider the effect that D has on four unit vectors lying along the coordinate axes of a two dimensional space. That is, we form the products

$$D\mathbf{x}_1 = \begin{pmatrix} 2 & 0 \\ 0 & 2 \end{pmatrix}\begin{pmatrix} 1 \\ 0 \end{pmatrix} = \begin{pmatrix} 2 \\ 0 \end{pmatrix}; \quad D\mathbf{x}_2 = \begin{pmatrix} 2 & 0 \\ 0 & 2 \end{pmatrix}\begin{pmatrix} 0 \\ 1 \end{pmatrix} = \begin{pmatrix} 0 \\ 2 \end{pmatrix}$$

$$D\mathbf{x}_3 = \begin{pmatrix} 2 & 0 \\ 0 & 2 \end{pmatrix}\begin{pmatrix} -1 \\ 0 \end{pmatrix} = \begin{pmatrix} -2 \\ 0 \end{pmatrix}; \quad D\mathbf{x}_4 = \begin{pmatrix} 2 & 0 \\ 0 & 2 \end{pmatrix}\begin{pmatrix} 0 \\ -1 \end{pmatrix} = \begin{pmatrix} 0 \\ -2 \end{pmatrix}$$

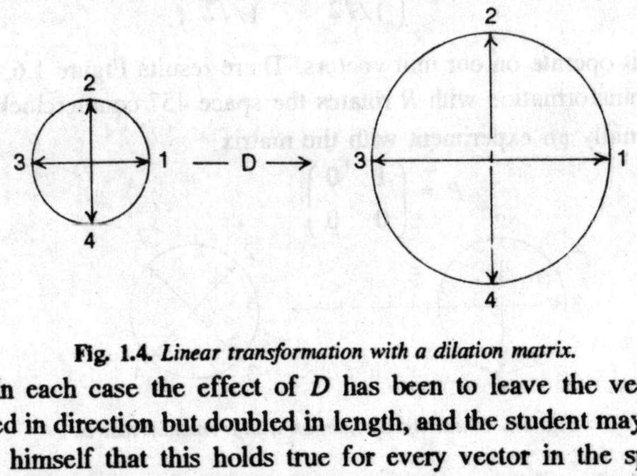

Fig. 1.4. *Linear transformation with a dilation matrix.*

In each case the effect of D has been to leave the vector unchanged in direction but doubled in length, and the student may readily satisfy himself that this holds true for every vector in the space, so that, for instance, the unit circle is sent into a circle of radius 2. Geometrically the picture is Figure 1.4, and we may reasonably describe D by saying that it is a dilation matrix, for it expands ("blows up") the space uniformly in every direction.

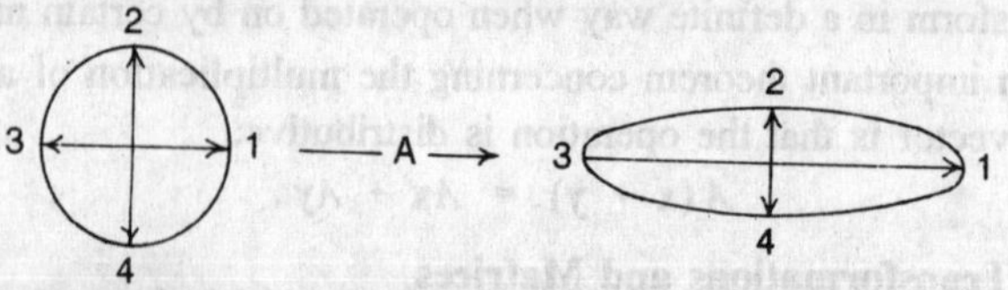

Fig. 1.5. *Linear transformation with a strain matrix.*

Now consider the matrix

$$A = \begin{pmatrix} 2 & 0 \\ 0 & \frac{1}{2} \end{pmatrix}$$

and let A operate on our four unit vectors $\mathbf{x}_1, \mathbf{x}_2, \mathbf{x}_3, \mathbf{x}_4$ and more generally on all unit vectors (whose termini must therefore lie on the unit circle). The student may verify that the resulting geometric picture is Figure 1. 5, so that vectors along the horizontal axis are doubled in length, those along the vertical axis are halved in length, and all other vectors are changed in both direction and length in such a way that circles concentric about the origin are sent into ellipses. Because the space is stretched in one direction and shrunken in another, A is an example of a strain matrix.

As another example, take

$$R = \begin{pmatrix} 1/\sqrt{2} & -1/\sqrt{2} \\ 1/\sqrt{2} & 1/\sqrt{2} \end{pmatrix}$$

and let it operate on our unit vectors. There results Figure 1.6, so that linear transformation with R rotates the space 45° counterclockwise.

Finally an experiment with the matrix

$$P = \begin{pmatrix} 1 & 0 \\ 0 & 0 \end{pmatrix}$$

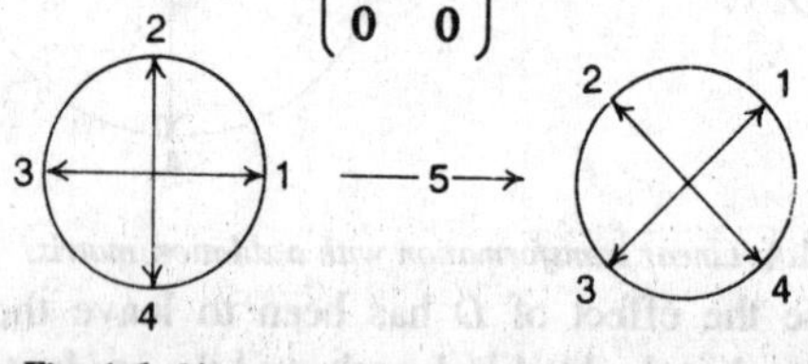

Fig. 1.6. *Linear transformation with a rotation matrix.*

will show that while $\mathbf{x}_1$ and $\mathbf{x}_3$ are reproduced identically, $\mathbf{x}_2$ and $\mathbf{x}_4$ are annihilated (sent into the null vector) as sketched in Figure. 1.7. P

evidently projects the entire space onto the horizontal axis.

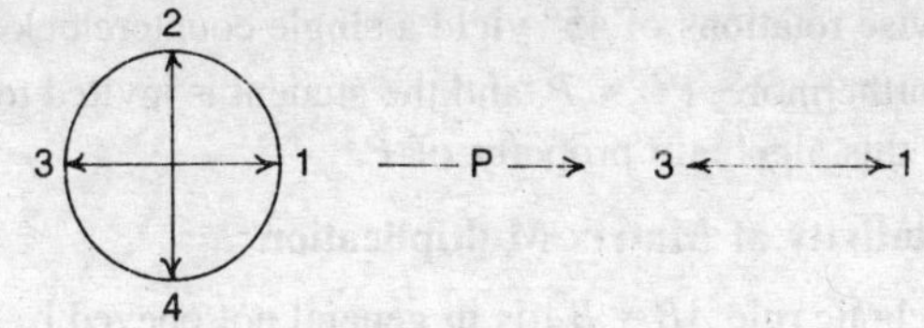

Fig. 1.7. *Linear transformation with a projection matrix.*

Successive Linear Transformations and Matrix Multiplication

Once a vector space has been transformed by a linear transformation, it may evidently be transformed again by another (or the same) linear transformation. Thus if $\mathbf{y} = A\mathbf{x}$, then $A\mathbf{y} = A(A\mathbf{x}) = A_2 x$. Geometrically this looks like Figure 1.8, in which

$$A^2 = \begin{pmatrix} 2 & 0 \\ 0 & \frac{1}{2} \end{pmatrix}\begin{pmatrix} 2 & 0 \\ 0 & \frac{1}{2} \end{pmatrix} = \begin{pmatrix} 4 & 0 \\ 0 & \frac{1}{4} \end{pmatrix}$$

The row-column rule for matrix multiplication was invented in order to reproduce algebraically these geometric transformations, so that the result of two linear transformations applied successively is another linear transformation. Another way of stating the same fact is that the product of two matrices is another matrix.

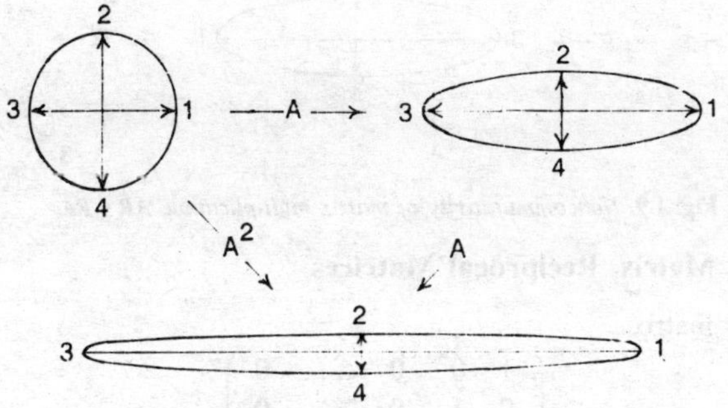

Fig. 1.8. *successive linear transformations.*

Similarly

$$R^2 = \begin{pmatrix} 1/\sqrt{2} & -1/\sqrt{2} \\ 1/\sqrt{2} & 1/\sqrt{2} \end{pmatrix}\begin{pmatrix} 1/\sqrt{2} & -1/\sqrt{2} \\ 1/\sqrt{2} & 1/\sqrt{2} \end{pmatrix} = \begin{pmatrix} 0 & -1 \\ 1 & 0 \end{pmatrix}$$

and the student may varify by experiment that this states that two counterclockwise rotations of 45° yield a single counterclockwise rotation of 90°. Furthermore, $P^2 = P$, and the student is invited to interpret geometrically this algebraic property of P.

The Commutativity of Matrix Multiplication

The algebraic rule $AB = BA$ is in general not obeyed by matrices, which is to say that matrix multiplication is not commutative. The student will find by actual calculation on the matrices Section 1.8 that $AR \neq RA$, and geometrically this states that the order of application of linear transformations makes a difference in the final result. In the example chosen, if we first stretch the space along the horizontal axis and then rotate it 45° counterclockwise, we obtain a different result than if we first rotated and then stretched along the horizontal axis (Figure 1.9).

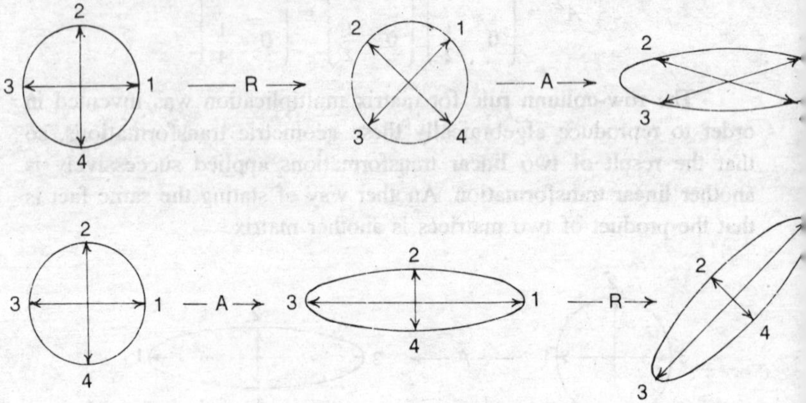

Fig. 1.9. *Noncommutativity of matrix multiplication, $AR \neq RA$.*

The Unit Matrix, Reciprocal Matrices

The matrix

$$\begin{pmatrix} 1 & 0 & 0 & \dots & 0 \\ 0 & 1 & 0 & \dots & 0 \\ \cdot & & & & \cdot \\ \cdot & & & & \cdot \\ \cdot & & & & \cdot \\ 0 & 0 & \dots & \dots & 1 \end{pmatrix}$$

which has unity along the *main diagonal* and zeros elsewhere is known as the *unit* or identity matrix. It is evidently a linear transformation that maps each vector in a vector space into itself, leaving the space unaltered. The student should check for himself the truth of the following theorems: $I\mathbf{x} = \mathbf{x}$ for every vector $\mathbf{x}$, and $IA = AI = A$ for every matrix A.

If two matrices H and G bear the relationship $GH = I$, they are said to be *reciprocal* to each other and they are written $H = G^{-1}$ or $G = H^{-1}$. Alternatively G^{-1} is said to be the *inverse* of G. An example is

$$AA^{-1} = \begin{pmatrix} 2 & 0 \\ 0 & \frac{1}{2} \end{pmatrix} \begin{pmatrix} \frac{1}{2} & 0 \\ 0 & 2 \end{pmatrix} = \begin{pmatrix} 1 & 0 \\ 0 & 1 \end{pmatrix} = I$$

The linear transformation A^{-1} evidently undoes what A does - A^{-1} shrinks along the horizontal axis and stretches along the vertical axis to just a sufficient degree to cancel the effect of A. Similarly

$$RR^{-1} = \begin{pmatrix} 1/\sqrt{2} & -1/\sqrt{2} \\ 1/\sqrt{2} & 1/\sqrt{2} \end{pmatrix} \begin{pmatrix} 1/\sqrt{2} & 1/\sqrt{2} \\ -1/\sqrt{2} & 1/\sqrt{2} \end{pmatrix} = \begin{pmatrix} 1 & 0 \\ 0 & 1 \end{pmatrix} = I$$

Not every matrix, however, possesses an inverse. The matrix P is an example of this, and in general a linear transformation which involves a projection cannot be inverted. The geometric interpretation of this is that a *singular* linear transformation (one that does *not* possess an inverse) maps many vectors in the original space into some subspace. This process cannot be inverted uniquely. On the other hand, nonsingular linear transformations (those matrices that possess an inverse) map the vector space onto itself in a one to one fashion, and this process can be inverted.

Some theorems concerning reciprocal matrices are the following:

1. $AA^{-1} = A^{-1}A = I$.
2. $(A^{-1})^{-1} = A$.
3. The inverse of a matrix, if it exists, is unique.

I have not given any method of computing the inverse of a matrix, for we shall not require it in our work. The problem is identical with finding a general solution to a set of n linear equations in n unknowns, and many methods are known for computing this solution (Cramer's rule, for example). A convenient test for the existence of an inverse is to examine the determinant of the matrix formed by treating the rows and columns of the matrix as the components of a determinant. If this determinant is zero the matrix is singular. If it is not zero, the matrix is nonsingular.

The Theory of Linear Dependence

The problem frequently arises in practice of expressing a given vector **y** as a *linear combination* of *n* given vectors $\mathbf{x}_j$

$$\mathbf{y} = c_1\mathbf{x}_1 + c_2\mathbf{x}_2 + \dots + c_n\mathbf{x}_n = \sum_1^n c_j\mathbf{x}_j$$

where the c_j are scalar constants to be determined. This single vector equation is equivalent to *n* ordinary scalar equations, which, when written out in detail, take the form

$$\begin{aligned} y_1 &= c_1x_{11} + c_2x_{12} + \dots + c_nx_{1n} \\ y_2 &= c_1x_{21} + c_2x_{22} + \dots + c_nx_{2n} \\ &\vdots \\ y_n &= c_1x_{n1} + c_2x_{n2} + \dots + c_nx_n \end{aligned}$$

in which

$$\begin{aligned} \mathbf{x}_1 &= (x_{11}, x_{21}, \dots, x_{n1}) \\ \mathbf{x}_2 &= (x_{12}, x_{22}, \dots, x_{n2}) \\ &\dots \end{aligned}$$

Our problem is: given the x_{ij} and the y_i, find the c_j. Stated in this way we see that we are asked to find the solution to a set of *n* linear equations in *n* unknowns. Such a set of equations has a unique solution only if the determinant of coefficients $|x_{ij}|$ does not vanish, and in general a set of *n* vectors in *n* dimensional space is said to be *linearly independent* if their components when assembled into the rows or columns of a determinant do not yield a zero determinant. Otherwise the vectors are said to be *linearly dependent*.

The geometric meaning of linear dependence is illustrated by an example: test the linear dependence of the vector sets

(a) $\mathbf{x}_1 = (1, -2)$ (b) $\mathbf{x}_1 = (1, -3)$

$\mathbf{x}_2 = (-2, 4)$ $\mathbf{x}_2 = (1, 2)$

We construct the determinants

(a) $\begin{vmatrix} 1 & -2 \\ -2 & 4 \end{vmatrix} = 0$ (b) $\begin{vmatrix} 1 & -3 \\ 1 & 2 \end{vmatrix} = 5$

whence it follows that (a) is linearly dependent and (b) is linearly independent. The student will find it suggestive to draw diagrams of each vector pair (a) and (b) after the manner of Figures 1.1 and 1.2. The vector set (a) will be observed to be collinear while the vectors of set (b) meet at an obtuse angle. Generally speaking, linearly dependent sets of vectors in two dimensions are collinear; linearly dependent sets of vectors in three dimensions are coplanar; and so on. A linearly independent set of vectors completely spans a vector space without being localized to some subspace of a lower number of dimensions.

Bases and Coordinate Systems

We have just seen that for a set of n linearly independent x_j, a unique expansion exists for any arbitrary vector y:

$$\mathbf{y} = c_1\mathbf{x}_1 + \dots + c_n\mathbf{x}_n$$

If the $\mathbf{x}_j$ are linearly dependent, the expansion either does not exist or is not unique. A set of n linearly independent vectors may thus be used as a *basis* or a *coordinate system* in a space of n dimensions in the sense that any arbitrary vector in the space may be expressed as a linear combination of the basis vectors in one and only one way. This is frequently a practical thing to do, and we have the theorem that if the $\mathbf{x}_j$ are linearly independent, then

$$c_1\mathbf{x}_1 + c_2\mathbf{x}_2 + \dots + c_n\mathbf{x}_n = b_1\mathbf{x}_1 + b_2\mathbf{x}_2 + \dots + b_n\mathbf{x}_n$$

implies $c_1 = b_1$; $c_2 = b_2$; . . . ; $c_n = b_n$.

Orthonormal Bases

From a purely mathematical point of view, no one basis has any preference over any other, and relations between vectors can be expressed just as well in all coordinate systems. From the point of view of the investigator interested in results, however, *orthonormal* bases have a distinct advantage, for the basis vectors can be manipulated very simply. An orthonormal set of vectors $\mathbf{E}_1, \mathbf{E}_2, \dots, \mathbf{E}_n$ has the property

$$\mathbf{E}_i \cdot \mathbf{E}_i = \delta_{ij} = 1 \quad \text{if } i = j$$
$$= 0 \text{ otherwise}$$

The symbol δ_{ij} is called the Kronecker delta. An example of an orthonormal coordinate system is

$$\mathbf{e}_1 = (1, 0, 0, \dots, 0)$$

$e_2 = (0, 1, 0, \ldots, 0)$

$e_n = (0, 0, 0, \ldots, 1)$

These are the usual Cartesian axes in which we conventionally express vectors. The set $\mathbf{e}_j$ is linearly independent

$$|e_{ij}| = |\delta_{ij}| = 1 \neq 0$$

and orthonormal $\mathbf{e}_i \cdot \mathbf{e}_j = \delta_{ij}$. Other orthonormal bases exist, however, and for example in two dimensions the set

$$\begin{aligned} \mathbf{E}_1 &= (1/\sqrt{2}, \quad 1/\sqrt{2}) \\ \mathbf{E}_2 &= (1/\sqrt{2}, \ -1/\sqrt{2}) \end{aligned} \qquad \ldots(1.1)$$

is linearly independent and orthonormal:

$$\begin{aligned} \mathbf{E}_1 \cdot \mathbf{E}_1 &= \mathbf{E}_2 \cdot \mathbf{E}_2 = 1 \\ \mathbf{E}_1 \cdot \mathbf{E}_2 &= \mathbf{E}_2 \cdot \mathbf{E}_1 = 0 \end{aligned}$$

Expansion in Orthonormal Bases

The determination of the unknown coefficients c_j in the expansion of a known vector **y** in terms of a given orthonormal basis is particularly simple:

$$\mathbf{y} = c_1\mathbf{E}_1 + c_2\mathbf{E}_2 + \ldots\ldots + c_n\mathbf{E}_n$$

If the E_j were not orthonormal, the problem, as we have seen, would be that of the solution of n linear equations in n unknowns. For the E_{ij} orthonormal, however, take the scalar product of each side of the above equation by E_1. The result is

$$\mathbf{E}_1 \cdot \mathrm{y} = c_1\mathbf{E}_1 \cdot \mathbf{E}_1 + 0 + 0 + \ldots + 0 = c_1$$

The method is quite general, and we may write $c_i = \mathbf{E}_j \cdot \mathbf{y}$.

In two dimensions a calculation using the orthonormal basis (1.1) is typical. Let $\mathbf{y} = (1, -2)$ and find c_1, c_2 for which

$$\mathbf{y} = c_1\mathbf{E}_1 + c_2\mathbf{E}_2$$

We have $c_1 = \mathbf{E}_1 \cdot \mathrm{y} = -1/\sqrt{2}$; $c_2 = \mathbf{E}_2 \cdot \mathrm{y} = 3/\sqrt{2}$, so that

$$\mathbf{y} = (-1/\sqrt{2})\mathbf{E}_1 + (3/\sqrt{2})\mathbf{E}_2$$

If the student will substitute from (1.1) into this formula he will find that the vector **y** is reproduced identically.

Transformation Theory

Imagine that we have some physical property, such as velocity or force, which can be represented by a vector in ordinary three dimensional space. It is apparent that depending on how we draw our set of orthogonal axes (or even if we use nonorthogonal axes) the mathematical representation of the physical vector (but not the vector itself) will change. Such changes do not, of course, have any effect on the physical property represented by the vector but only reflect the various mathematical choices open to the investigator. Because the mathematician has this freedom of choice in the way he can describe a given physical situation, he quite properly makes choices in such a way as to simplify the algebraic or computational details of his work. In other words, he often needs a method of calculating the components of a vector in a new coordinate system when he is given its components in some other coordinate system. This calculation can be effected by matrix algebra, and it is shown in texts upon the subject that if the components of a vector in the old coordinate system are $\mathbf{x}$ and the components of the same (physical) vector in the new coordinate system are $\mathbf{x}'$, then the numbers $\mathbf{x}'$ are related to the numbers $\mathbf{x}$ by $\mathbf{x}' = M\mathbf{x}$ where M is the *matrix of the transformation.* Of particular importance in practical work is the *orthogonal transformation* by means of which one set of orthogonal axes is converted into another. Intuitively we should expect an orthogonal transformation in two or three dimensional space to correspond to a rotation or a reflection of coordinate axes, and it is the generalization of this expectation to spaces of a higher number of dimensions which has led to the n dimensional *orthogonal matrix.* If a matrix M is orthogonal, it has the remarkable property that its n rows and its n columns each separately constitute a set of n orthonormal vectors.

We shall consider further properties of these matrices in due course; for now it is sufficient to accept the idea that if $\mathbf{E}_1, \mathbf{E}_2, \ldots, \mathbf{E}_n$ are orthonormal, then for M an orthogonal matrix, the vectors $\varepsilon_1 = M\mathbf{E}_1, \varepsilon_2 = M\mathbf{E}_2, \ldots, \varepsilon_n = M\mathbf{E}_n$ also constitute an orthonormal set.

The Eigenvalue-Eigenvector Problem

If a vector $\mathbf{x}$ has the property with respect to a give matrix K of satisfying

$$K\mathbf{x} = \lambda\mathbf{x}$$

where λ is a scalar, $\mathbf{x}$ is said to be an *eigenvector* of K with *eigenvalue* λ. Geometrically, *if a vector space undergoes a linear transformation, those vectors that are altered in length but not in direction are eigenvectors of that linear transformation. This statement must be interpreted to include "negative" length—that is, the vector may be reflected backward through the origin and still be an eigenvector, in this case with a negative eigenvalue.* Some examples:

1. Given $A = \begin{pmatrix} 2 & 0 \\ 0 & \frac{1}{2} \end{pmatrix}$. Then $A\mathbf{x} = \lambda\mathbf{x}$ is satisfied if $\mathbf{x} = \begin{pmatrix} 1 \\ 0 \end{pmatrix}$ and $\lambda = 2$, or if $\mathbf{x} = \begin{pmatrix} 0 \\ 1 \end{pmatrix}$ and $\lambda = \frac{1}{2}$.

2. The rotation matrix R of Section 1.8 possesses no nonnull eigenvectors or eigenvalues of the type we have been studying that is, whose components belong to the field of real numbers.

3. The projection matrix P of Section 1.8 possesses eigenvectors of the form $\mathbf{x} = \begin{pmatrix} x \\ 0 \end{pmatrix}$ with arbitrary x, all of which have eigenvalue $\lambda = 1$.

4. Every vector in n dimensional space is an eigenvector of the identity matrix I, all with eigenvalue $\lambda = 1$.

The Secular Equation

In quantum mechanics, the eigenvalue-eigenvector problem presents itself in the following way: given a matrix K, find its eigenvalues and eigenvectors. In matrix terminology we are to find all λ's and $\mathbf{x}$'s that satisfy $K\mathbf{x} = \lambda\mathbf{x}$. Written out in full, this vector equation is equivalent to n scalar equations:

$$\begin{aligned} k_{11}x_1 + k_{12}x_2 \ldots + k_{1n}x_n &= \lambda x_1 \\ \vdots \qquad & \qquad \vdots \\ k_{n1}x_1 + k_{n2}x_2 \ldots + k_{nn}x_n &= \lambda x_n \end{aligned}$$

which may be rearranged to

$$\begin{aligned} (k_{11} - \lambda)x_1 + k_{12}x_2 \ldots + k_{1n}x_n &= 0 \\ k_{21}x_1 + (k_{22} - \lambda)x_2 \ldots + k_{2n}x_n &= 0 \\ \vdots \qquad & \qquad \vdots \\ k_{n1}x_1 + \qquad \ldots + (k_{nn} - \lambda)x_n &= 0 \end{aligned}$$

This is a set of n homogeneous, linear equations in n unknowns, and such a set always has the solution $x_1 = x_2 = \ldots = x_n = 0$ no matter what the value of λ. The null vector is thus an eigenvector of every matrix with arbitrary eigenvalue. In practice, however, we are never interested in this solution, called the *trivial solution*. We always require solutions other than the trivial one, and by a well known theorem these will exist only if λ is chosen so as to set the determinant of coefficients of these equations equal to zero:

$$[k_{ij} - \lambda\delta_{ij}] = 0 \quad or \quad [K - \lambda I] = 0$$

This determinantal equation is called the *secular equation* of the eigenvalue problem and serves to determine the eigenvalues λ. When the secular determinant $[K - \lambda I]$ is expanded, it turns out to be a polynomial of degree n in λ called the *characteristic polynomial.* Because a polynomial of degree n can have at most n distinct roots, there are at most n distinct eigenvalues $\lambda_1, \lambda_2, \ldots \lambda_n$ which are roots of the characteristic polynomial.

Some examples : (1) Find the eigenvalues of

$$A = \begin{pmatrix} 2 & 0 \\ 0 & \frac{1}{2} \end{pmatrix}$$

We construct the determinant $|A - \lambda I|$ and set it equal to zero :

$$\begin{vmatrix} 2 - \lambda & 0 \\ 0 & \frac{1}{2} - \lambda \end{vmatrix} = (2 - \lambda)(\tfrac{1}{2} - \lambda) = 0$$

The polynomial $(2 - \lambda)(\frac{1}{2} - \lambda) = \lambda^2 - (\frac{5}{2})\lambda + 1$ is evidently the characteristic polynomial with roots $\lambda_1 = 2$ and $\lambda_2 = \frac{1}{2}$ as eigenvalues.

(2) Find the eigenvalues of

$$K = \begin{pmatrix} 6 & 2 \\ 2 & 3 \end{pmatrix}$$

Construct

$$|K - \lambda I| = \begin{vmatrix} 6 - \lambda & 2 \\ 2 & 3 - \lambda \end{vmatrix} = \lambda^2 - 9\lambda + 14 = (\lambda - 2)(\lambda - 7) = 0$$

Hence $\lambda_1 = 7$; $\lambda_2 = 2$.

In the examples above the eigenvalues found were all real numbers and they were distinct, with no repeated roots. Matrices can have complex numbers as eigenvalues, but such matrices will not occur in our work. Matrices with repeated eigenvalues, that is, with characteristic polynomials that have multiple roots do, however occur. Such repeated eigenvalues are said to be *degenerate*.

Eigenvectors

To each root λ_i of the characteristic polynomial there corresponds an eigenvector. To find the components of the eigenvector corresponding to, say, λ_1, put $\lambda = \lambda_1$ in our set of homogeneous, linear equations $K\mathbf{x} = \lambda_1\mathbf{x}$ and solve for the components of **x**. For the matrix

$$A = \begin{pmatrix} 2 & 0 \\ 0 & \frac{1}{2} \end{pmatrix}$$

the calculation is

$$(2 - \lambda_1)\,x_1 + (0)\,x_2 = 0$$

$$(0)\,x_1 + (\tfrac{1}{2} - \lambda_1)\,x_2 = 0$$

or, since $\lambda_1 = 2$,

$$(0)\,x_1 + (0)\,x_2 = 0$$

$$(0)\,x_1 + (-\tfrac{3}{2})\,x_2 = 0$$

It follows from these equations that they are satisfied by $x_2 = 0$ and $x_1 =$ anything. The result is typical, and it can be shown from the general theory of linear equations that the eigenvector corresponding to a nondegenerate eigenvalue is determined only to within an aarbitrary component; hence if one component is fixed arbitrarily, the others are uniquely determined. Geometrically, the eigenvector corresponding to a nondegenerate eigenvalue is fixed as to its direction but not as to its length. Because the length is arbitrary, it is conventional (but not essential) to normalize the eigenvector to unit length. Thus in our example normalization requires $x_1^2 + x_2^2 = 1$; but $x_2 = 0$, so that $x_1{}^2 = 1$, and I take the positive square root $x_1 = 1$, although $x_1 = -1$ would be just as good. The complete, normalized eigenvector corresponding to $\lambda_1 = 2$ is therefore

$$\mathbf{x}_1 = \begin{pmatrix} 1 \\ 0 \end{pmatrix}$$

In a similar way one finds for the eivenvalue $\lambda_2 = \frac{1}{2}$ a normalized eigenvector

$$x_2 = \begin{pmatrix} 0 \\ 1 \end{pmatrix}$$

The student should turn back to Figure 1.5 and locate these eigenvectors in the linear transformation performed by A on the vector space of two dimensions.

Another example:

$$K = \begin{pmatrix} 6 & 2 \\ 2 & 3 \end{pmatrix}; \ \lambda_1 = 7; \ \lambda_2 = 2$$

For $\lambda = \lambda_1$,

$$(6 - 7)x_1 + 2x_2 = 0$$

$$2x_1 + (3 - 7)x_2 = 0$$

which reduces to a single equation :

$$-x_1 + 2x_2 = 0$$

If we want the eigenvector to be normalized, we must also have $x_1{}^2 + x_2{}^2 = 1$, so that the resulting eigenvector is

$$\mathbf{x}_1 = \begin{pmatrix} 2/\sqrt{5} \\ 1/\sqrt{5} \end{pmatrix}$$

Similarly for $\lambda_2 = 2$, we have a normalized eigenvector

$$\mathbf{x}_2 = \begin{pmatrix} -1/\sqrt{5} \\ 2/\sqrt{5} \end{pmatrix}$$

It is extremely suggestive to follow these operations geometrically. Consider Figure 1.10. Before transformation of the space by K, the eigenvectors $\mathbf{x}_1$ and $\mathbf{x}_2$ are of unit length. After transformation with K, $\mathbf{x}_1$ becomes $7\mathbf{x}_1$ and $\mathbf{x}_2$ becomes $2\mathbf{x}_2$. If the student experiments with other unit vectors, he will find that the unit circle is transformed by K into an ellipse with principal axes oriented along the directions x_1 and x_2.

Degenerate Eigenvalues

We have just observed that the eigenvectors of a nondegenerate eigenvalue are determined to within an arbitrary component. Geometrically, this means that we know the line through the origin on which

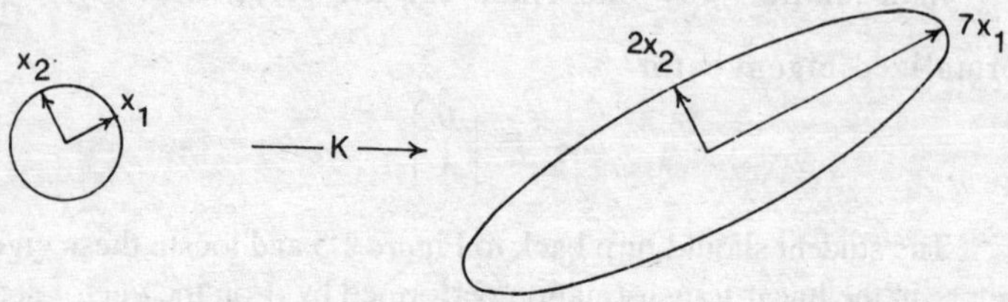

Fig. 1.10. *A linear transformation showing eigenvectors.*

the vector lies, and any vector on this line is an eigenvector. It can be shown from the theory of linear equations that if an eigenvalue is doubly degenerate, two of the components of its eigenvector may be chosen arbitrarily; geometrically this means that the eigenvector is known to lie in a certain subspace of two dimensions, and any vector in this subspace is an eigenvector, all of them belonging to the same eigenvalue. Similarly the eigenvectors belonging to a triply degenerate eigenvalue all lie within a three dimensional subspace, and so on.

To illustrate these properties, the dilation matrix

$$D = \begin{pmatrix} 2 & 0 \\ 0 & 2 \end{pmatrix}$$

is doubly degenerate, for the secular equation

$$\begin{vmatrix} 2-\lambda & 0 \\ 0 & 2-\lambda \end{vmatrix} = (2-\lambda)^2 = 0$$

has root $\lambda = 2, 2$. When we try to find the components of an eigenvector for $\lambda = 2$, we have to solve the linear equations

$$(2-\lambda)x_1 = 0$$

$$(2-\lambda)x_2 = 0$$

which for $\lambda = 2$ tell us only that $\mathbf{x}_1$ and $\mathbf{x}_2$ are arbitrary. But this makes very good geometric sense, for Figure 1.4 shows that *all* vectors in the x_1, x_2 plane are sent into multiples of themselves when operated on by D. This is just another way of saying that all vectors in the x_1, x_2 plane are eigenvectors of D.

Transposed Matrices

The matrix K^T which is formed from K by rearranging the columns of K into the rows of K^T is known as the *transpose* of K. Symbolically $(k_{ij})^T = (k_{ji})$. For example, the transpose of

$$\begin{pmatrix} 1 & 2 & 3 \\ 0 & -1 & 4 \\ 0 & 0 & 6 \end{pmatrix}$$

$$\begin{pmatrix} 1 & 0 & 0 \\ 2 & -1 & 0 \\ 3 & 4 & 6 \end{pmatrix}$$

Up to now we have not specified in our notation for vectors whether **x** is to be written as a row or a column. We henceforth specify **x** to be a colunm and $\mathbf{x}^T$ to be the same rector written as row. Thus $\mathbf{x}^T$ is the transpose of **x**. In terms of this notation and the ordinary rules of matrix multiplication, the scalar product of two vectors **x** and **y** may be written $\mathbf{x} \cdot \mathbf{y} = \mathbf{x}^T\mathbf{y}$.

An important thorem concerning transposed matrices states that the transpose of a product is the product of the transposes taken in reverse order. This theorem is true for the scalar product of two vectors, for the product of a matrix times a vector, and for the product of a matrix times a matrix. Symbolically,

$$\mathbf{x}^T\mathbf{y} = \mathbf{y}^T\mathbf{x}\,; \quad (G\mathbf{x})^T = \mathbf{x}^T G^T; \quad (AB)^T = B^T A^T \qquad ...(1.2)$$

As an example of the second of these, if

$$G = \begin{pmatrix} 1 & 0 \\ 1 & 3 \end{pmatrix} \text{ and } \mathbf{x} = \begin{pmatrix} 1 \\ 0 \end{pmatrix}$$

then

$$G\mathbf{x} = \begin{pmatrix} 1 & 0 \\ 1 & 3 \end{pmatrix}\begin{pmatrix} 1 \\ 0 \end{pmatrix} = \begin{pmatrix} 1 \\ 1 \end{pmatrix}$$

$$\mathbf{x}^T G^T = (1\,,0)\begin{pmatrix} 1 & 1 \\ 0 & 3 \end{pmatrix} = (1, 1) = (G\mathbf{x})^T$$

A matrix K for which $K = K^T$ is said to be *symmetric.* Such matrices have their components symmetrically disposed with respect to the main diagonal, and they will play a role of peculiar importance in our work. The matrix A of Section 1.8 and the matrix K of Section 1.19 are examples of symmetric matrices.

The orthogonal matrix introduced in Section 1.16 has the property that its transpose is equal to its inverse $M^T = M^{-1}$ or $M^TM = I$, as the student may check by experimentation with the matrix R of Section 1.8.

Eigenvectors of a Symmetric Matrix

A very important theorem states that the eigenvectors of a real, symmetric matrix which are associated with distinct eigenvalues are orthogonal. *Proof*: Let K be a real, symmetric matrix with two distinct eigenvalues λ_1 and λ_2 and let the eigenvectors associated with these eigenvalues be Ψ_1 and Ψ_2. Then

$$k\Psi_1 = \lambda_1\Psi_1$$
$$k\Psi_2 = \lambda_2\Psi_2$$

with $\lambda_1 \neq \lambda_2$. From the scalar product of the first equation with Ψ_2 and the scalar product of the second equation with Ψ_1.

$$\Psi_2^T K\Psi_1 = \lambda_1\Psi_2^T\Psi_1$$
$$\Psi_1^T K\Psi_2 = \lambda_2\Psi_1^T\Psi_2 \quad ...(1.3)$$

Now transpose both sides of the second part of (1.3), using the rule (1.2) for the transpose of a product, and subtract the result from the first of equations 1.3:

$$\psi_2^T K\psi_1 - \psi_2{}^T K^T \psi_1 = \lambda_1\psi_2{}^T\psi_1 - \lambda_2\psi_2{}^T\psi_1$$

or

$$\Psi_2^T (K - K^T)\Psi_1 = (\lambda_1 - \lambda_2)\Psi_2^T\Psi_1$$

Now if $K = K^T$, then the left hand side is zero, and there remains

$$(\lambda_1 - \lambda_2)\Psi_2^T\Psi_1 = 0$$

But we are given that $\lambda_1 \neq \lambda_2$, and it follows that we must have $\Psi_2{}^T\Psi_1 = 0$, meaning that Ψ_1 and Ψ_2 are orthogonal. Note that this is not necessarily the case if both Ψ_1 and Ψ_2 belong to a degenerate eigenvalue $\lambda_1 = \lambda_2$.

Eigenvalues of a Symmetric Matrix

Another very important theorem states that the eigenvalues of a real, symmetric matrix are all real. To prove this, let the operation of taking the complex conjugate of a scalar or a vector be denoted by an asterisk (*), so that λ^* is the complex conjugate of λ and Ψ^* is the complex conjugate of Ψ. Then

$$K\Psi = \lambda\Psi$$

and if we take the complex conjugate of each side of this equation

$$K\Psi^* = \lambda^*\Psi^*$$

where K, being defined real, is unaffected by taking its complex conjugate. Now from our previous theorem, if $\lambda \neq \lambda^*$, then Ψ^* must be orthogonal to Ψ. With the exception of the trivial solution $\Psi = 0$, however, it is impossible for a vector to be orthogonal to its own complex conjugate (try it !), and it must follow that $\lambda = \lambda^*$, which is to say that λ is real.

Recapitulation

We can summarize the importance of these theorems by noting that if all of the eigenvalues of a real, symmetric matrix are distinct, there will be n in number and to each one is assigned as eigenvector that is orthogonal to every other eigenvector. If we by convention normalize these eigenvectors, then they form an orthonormal set in the space of n dimensions and are thus a suitable basis (or coordinate system) in the space of n dimensions. Hence whenever a symmetric matrix K occurs in a physical problem, the computational details of that problem usually become enormously simplified if the coordinate axes are chosen to lie along the eigenvectors of K. If the eigenvalues are not all distinct, they may be grouped into degenerate sets, and every eigenvector in a given degenerate set will be automatically orthogonal to every eigenvector belonging to another degenerate set. Within the members of a given degenerate set, however, the eigenvectors are not necessarily orthogonal. We have seen that they all lie within some subspace of the vector space of n dimensions, the dimension of the subspace being equal to the degeneracy of the eigenvalue, so that the eigenvectors of a doubly degenerate eigenvalue lie within a plane. The eigenvectors of a triply degenerate eigenvalue lie within a subspace (volume) of three dimensions, and so on.

So long an eigenvector belonging to a degenerate eigenvalue lies within its assigned subspace, its orientation with that subspace is arbitrary. It is usually convenient, however, to choose these degenerate eigenvectors to be orthonormal, so that they form a coordinate system within the assigned subspace. No set of orthonormal, degenerate eigenvectors is, however, to be preferred over any other *ab initio*; which set is chosen usually depends on other features of the problem in hand.

In any case, once subsets of orthonormal axes have been chosen for every degenerate eigenvalue, the entire vector space of n dimensions is spanned by n orthonormal eigenvectors, and what I said previously still stands: the eigenvectors of a real, symmetric matrix may be taken to be a basis in the vector space of n dimensions.

A Degenerate Eigenvalue Problem

Let us analyze in detail a specific example. Find the eigenvalues and a set of normalized eigenvectors Ψ_k for

$$K = \begin{pmatrix} 5 & \sqrt{2} & 1 \\ \sqrt{2} & 6 & \sqrt{2} \\ 1 & \sqrt{2} & 5 \end{pmatrix} = K^T$$

The characteristic polynomial $f(\lambda) = |K - \lambda I|$ turns out to be

$$-f(\lambda) = \lambda^3 - 16\lambda^2 + 80\lambda - 128$$
$$= (\lambda) - 4)^2 (\lambda - 8)$$

Hence the eigenvalues are $\lambda_1 = \lambda_2 = 4$; $\lambda_3 = 8$. The third normalized eigenvector corresponding to the nondegenerate λ_3 is easily obtained. Its components satisfy

$$(5 - 8)x_1 + (\sqrt{2})x_2 + (1)x_3 = 0$$
$$(\sqrt{2})x_1 + (6 - 8)x_2 + (\sqrt{2})x_3 = 0$$
$$x_1{}^2 + x_2{}^2 + x2^2 = 1$$

whence $\Psi_3 = (\frac{1}{2}, \sqrt{2}/2, \frac{1}{2})$. For$\lambda_1 = \lambda_2 = 4$, however, the set of three homogeneous, linear equations reduces to a single equation

$$x_1 + \sqrt{2}x_2 + x_3 = 0 \qquad \text{...(1.4)}$$

and both eigenvectors satisfy this equation together with the normalization condition

$$x_1{}^2 + x_2{}^2 + x_3{}^3 = 1$$

If we pick any arbitrary value for one of the components, then the other two are determined; thus, for example, if $x_2 = 0$, then (1.4) becomes $x_1 + x_3 = 0$, and a possible eigenvector is $\Psi_1 = (\sqrt{2}/2, 0, -\sqrt{2}/2)$, which we note in passing, is orthogonal to Ψ_3. Another eigenvector Ψ_2 associated with the eigenvalue 4 is obtained by making its components satisfy equation 1.4, and for convenience we also choose Ψ_2 to be orthogonal to Ψ_1):

$$\Psi_2^T \Psi_1 = ((\sqrt{2}/2)x_1 - (\sqrt{2}/2)x_3 = 0$$

When normalized, the result is $\Psi_2 = (\frac{1}{2}, -\sqrt{2}/2, \frac{1}{2})$.

A clearer picture of the meaning of these operations is obtained if we construct the corresponding linear transformation in the vector space of three dimensions. If K operates on all unit vectors (those with their ends resting on the unit sphere), the eigenvector Ψ_3 is lengthened by a factor 8, and *all* unit vectors in the subspace spanned by Ψ_1, Ψ_2 are lengthened by a factor 4 (Figure 1.11). The unit sphere is thus elongated into an ellipsoid retaining its symmetry in the 1, 2 plane. Had our matrix been nondegenerate, the sphere would have been converted into an ellipsoid with all three of its principal axes distinct. The geometric meaning of the degenerate subspace spanned by the degenerate eigenvectors Ψ_1 and Ψ_2 is thus clear, for there is no unique way of drawing orthogonal axes within it; and the choice of a coordinate system is to this extent ideterminate.

The student should correlate these pictures with Figure 1.4 for the degenerate dilation matrix D, which possesses no unique principal

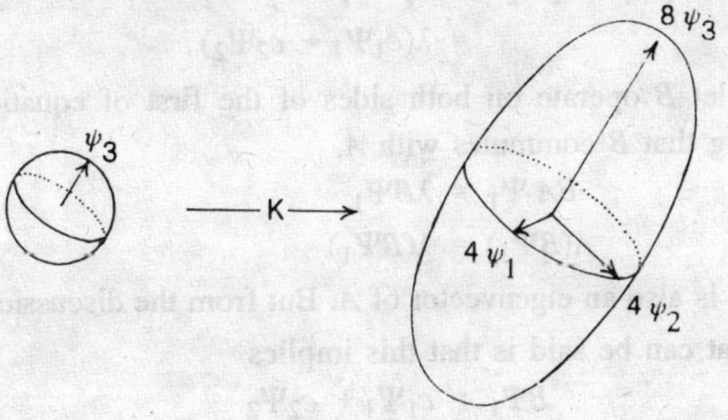

Fig. 1.11. *A degenerate linear transformation showing eigenvectors.*

axes after the fashion of matrix A of Figure 1.5. It is evident that the subspace spanned by the degenerate eigenvectors of a symmetric matrix is locally dilated by the linear transformation.

Eigenvectors of Commuting, Symmetric Matrices

If A and B commute, $AB = BA$, they possess a complete set of eigenvectors in common. Thus suppose that

$$A\Psi = \lambda'\Psi$$

Operate on both sides by B:

$$BA\Psi = \lambda B\Psi$$

But $BA = AB$, so that

$$AB\Psi = \lambda B\Psi$$

or

$$A(B\Psi) = \lambda(B\Psi)$$

This states that $B\Psi$ is an eigenvector of A. But Ψ is also an eigenvector of A, and if λ is nondegenerate, then $B\Psi$ can only be some multiple of Ψ,

$$B\Psi = \mu\Psi$$

which is to say that Ψ is also an eigenvector of B. This will be true for every nondegenerate eigenvector of A, and all such eigenvectors are therefore shared in common by A and B

If λ is doubly degenerate, then there exist two normalized, mutually orthogonal vectors Ψ_1 and Ψ_2 for which

$$A\Psi_1 = \lambda\Psi_1 ; \; A\Psi_2 = \lambda\Psi_2 \qquad ...(1.5)$$

We note further more from Section 1.20 and Problem 1.17 that any linear combination of Ψ_1 and Ψ_2 is also an eigenvector of A:

$$\begin{aligned} A(c_1\Psi_1 + c_2\Psi_2) &= c_1 A\Psi_1 + c_2 A\Psi_2 = c_1\lambda\Psi_2 + c_2\lambda\Psi_2 \\ &= \lambda(c_1\Psi_1 + c_2\Psi_2) \end{aligned}$$

Now let B operate on both sides of the first of equations 1.5, remembering that B commutes with A,

$$\begin{aligned} BA\,\Psi_1 &= \lambda B\Psi_1 \\ A(B\Psi_1) &= \lambda(B\Psi_1) \end{aligned}$$

so that $B\Psi_1$ is also an eigenvector of A. But from the discussion above the most that can be said is that this implies

$$B\Psi_1 = c_1\Psi_1 + c_2\Psi_2$$

with c_1 and c_2 unknown. Only for the special case $c_2 = 0$ could Ψ_1 be taken as an eigenvector of B.

On the other hand, the original choice of the basis pair Ψ_1, Ψ_2 was to an extent arbitrary; the question now is, could we by another choice of basis pair find a set of common eigenvectors for A and B ? That is, could we chosse c_1 and c_2 in such a way as to make not Ψ_1 and Ψ_2 individually, but the linear combination $c_1\Psi_1 + c_2\Psi_2$ an eigenvector of B ?

$$B(c_1\Psi_1 + c_2\Psi_2) = \mu(c_1\Psi_1 + c_2\Psi_2)\,? \qquad ...(1.6)$$

To answer this question we take the scalar products of both sides of equation 1.6 with Ψ_1 and with Ψ_2

$$\begin{aligned} c_1(\Psi_1^T B\Psi_1) + c_2(\Psi_1^T B\Psi_2) &= \mu c_1 \\ c_1(\Psi_2^T B\Psi_1) + c_2(\Psi_2^T B\Psi_2) &= \mu c_2 \end{aligned} \qquad ...(1.7)$$

in which the right hand side has been simplified by virtue of the orthonormality of Ψ_1 and Ψ_2. The four scalar quantities

$$H_{ij} = \Psi_i^T B\Psi_i \qquad ...(1.8)$$

are the components of the 2 × 2 matrix of coefficients of a pair (1.7) of linear equations in c_1 and c_2. From the rule on transposition of a product (equation 1.2)

$$(\Psi_1^T B\Psi_2)^T = \Psi_2^T B^T\Psi_1$$

But $B = B^T$ is symmetric, so that the matrix of coefficients (1.8) is also symmetric with $H_{12} = H_{21}$. Rearranging equations 1.7,

$$(H_{11} - \mu)c_1 + H_{12}c_2 = 0$$

$$H_{21}c_1 + (H_{22} - \mu)c_2 = 0$$

which is a set of two homogeneous, linear equations in two unknown c's. In the usual way, such a set can have a nontrivial solution only if the determinant of coefficients vanishes

$$\begin{vmatrix} H_{11} - \mu & H_{12} \\ H_{21} & H_2 - \mu \end{vmatrix} = 0 \qquad ...(1.9)$$

The root μ of this equation give two of the eigenvalues of B, its two normalized eigenvectors each give a pair of cofficients c_1, c_2 which will make the linear combination $c_1\Psi_1 + c_2\Psi_2$ a simultaneous, normalized eigenvector of both A and B.

If λ is n fold degenerate, the secular equation to be solved is of the nth order.

$$| H_{ij} - \mu\delta_{ij} | = 0$$

whose n roots μ belong to B and whose n orthonormal eigenvectors $(c_{k1}, c_{k2}, ..., c_{kn})$ constitute the sets of coefficients necessary to construct n linear combinations,

$$c_{k1}\Psi_1 + c_{k2}\Psi_2 + \ldots + c_{kn}\Psi_n$$

which are simultaneoulsy eigenvectors of both A and B.

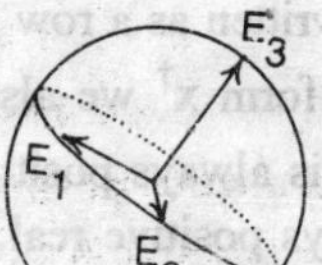

Fig. 1.12. *Eigenvectors before linear transformation.*

As in Section 1.25, a few geometric diagrams may help clarify the meaning of this theorem. Suppose that we have a nondegenerate matrix G with eigenvalues λ_k and eigenvectors $\mathbf{E}_k$ which commutes with the degenerate K of Section 1.25. Our theorem then states that G and K have eigenvectors in common. Because G is nondegenerate, the orientation of its egenvectors is uniquely determined, say as in Figure 1.12, so that according to our theorem, $\mathbf{E}_3$ automatically coincides with the nondegenerate Ψ_3. Let us picture the result of the linear transformation G on the unit sphere to be the flattened ellipsoid of Figure 1.13. By comparison of Figure 1.13 with Figure 1.11 for the linear transformation K, we perceive that while Ψ_1 and Ψ_2 do not necessarily coincide with $\mathbf{E}_1$ and $\mathbf{E}_2$, they can be made to do so by a

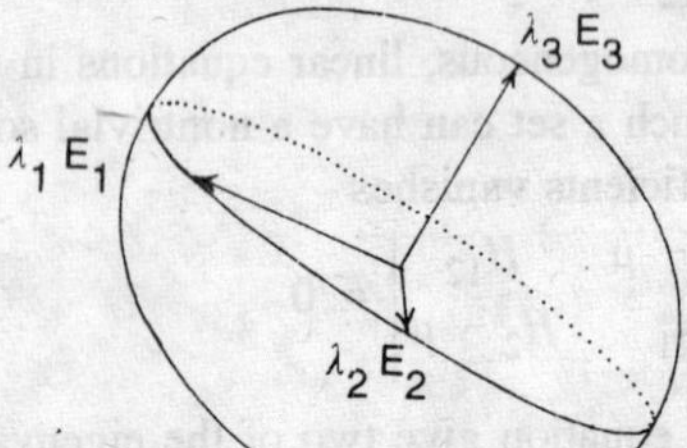

Fig. 1.13. *Eigenvectors after linear transformation.*

rotation of coordinate axes (a change of basis) in the K diagram. This is permissible, for any vector lying in the subspace 1, 2 of the K diagram is an eigenvector of K. It is the rotation of these axes (i.e., the construction of an orthogonal transformation), which is the result of the solution of the secular equation 1.9.

Vector Spaces with Complex Numbers

So far we have considered only vectors and matrices whose components are real numbers. All of our theorems generalize to the case where complex numbers are admitted. The scalar product of two vectors is written $\mathbf{x}^\dagger\mathbf{y}$, where $\mathbf{x}^\dagger$ is written as a row and $\mathbf{y}$ as a column, but in addition to transposing $\mathbf{x}$ to form $\mathbf{x}^\dagger$ we also take its complex conjugate. With this definition $\mathbf{x}^\dagger\mathbf{x}$ is always positive real, so that the norms of complex vectors are always positive real.

In a similar way the transposition of a matrix is always accompanied by taking the complex conjugate of every component. Thus for complex matrices, $K^\dagger$ means transposition of K accompanied by complex conjugation. If $K = K^\dagger$, the matrix is said to be *Hermitian*, which is a generalization of the idea of symmetry for real matrices. As for symmetric, real matrices, the following theorems exist for Hermitian matrices:

1. The eigenvalues of an Hermitian matrix are real.
2. The eigenvectors of an Hermitian matrix beloging to different eigenvalues are orthogonal.
3. Hermitian matrices that commute possess a complete set of eigenvectors in common.

The role of an orthogonal matrtix in real matrix theory is played in complex matrix theory by the *unitary* matrix, which has the property $M^\dagger = M^{-1}$, or $M^\dagger M = I$. The student should verify by actual calculation that for $i = \sqrt{-1}$ and

$$M = \begin{pmatrix} 1/\sqrt{2} & 1/\sqrt{2} \\ -i/\sqrt{2} & i/\sqrt{2} \end{pmatrix} \tag{1.10}$$

then

$$M^\dagger = \begin{pmatrix} 1/\sqrt{2} & i/\sqrt{2} \\ 1/\sqrt{2} & -i/\sqrt{2} \end{pmatrix}$$

and

$$M^\dagger M = \begin{pmatrix} 1/\sqrt{2} & i/\sqrt{2} \\ 1/\sqrt{2} & -i/\sqrt{2} \end{pmatrix} \begin{pmatrix} 1/\sqrt{2} & 1/\sqrt{2} \\ -i/\sqrt{2} & i/\sqrt{2} \end{pmatrix} = \begin{pmatrix} 1 & 0 \\ 0 & 1 \end{pmatrix}$$

THE FIRST POSTULATE OF WAVE MECHANICS

The ideas just discussed may be incorporated into a first postulate of wave mechanics. The postulate gives the relationship between the mathematical representation of a state by means of a wave function and the physical determination of the state by means of experiments or measurement. It may be formulated as follows.

Postulate **A**. The state of a mechanical system is represented by wave function $\Psi(q_1, q_2, q_3, \ldots, q_n, t)$. This is a function of the independent cartesian coordinates of the "particles" composing the system and of the time t. It represents the state of the system defined by postulates A_1, A_2 and A_3.

Postulate A_1 : $W(q_1, q_2, \ldots, q_n, t) = |\Psi(q_1, q_2, \ldots, q_n, t)|^2$ where $W\, dq_1 \ldots dq_n$ is the probability that a simultaneous measurement of the various coordinates will give results lying in the ranges q_1 to $(q_1 + dq_1)$, q_2 to $(q_2 = dq_2)$, ... q_n to $(q_n + dq_n)$ at the time t.

The probability distribution of the momenta is also given by the function Ψ through the following rule: Let

$$\phi(p_1, p_2, \ldots\ldots, p_n, t)$$

$$= h^{-\frac{1}{2}n} \int dq_1 \ldots dq_n \, e^{-(i/h)(p_i q_i + \ldots + p_n q_n)} \, \psi(q_1, \ldots q_n, t). \quad \ldots(1.11)$$

The momentum probability distribution is given directly in term or these functions.

Postulate A_2 : $W(p_1, p_2, \ldots p_n, t) = |\phi(p_1, \ldots, p_n, t)|^2$

where $W(p_1, p_2, \ldots\ldots p_n, t)$ is defined as is $W(q_1, \ldots\ldots q_n, t)$ above. It is important to note that although the knowledge of the state of the system, the knowledge of the function Ψ, suffices to determine the probability distributions of the coordinates and momenta, the converse is not true. The wave function contains more than these probability distributions, and these distributions do not determine the function uniquely.

Normalization of Wave Functions

In order that the Postulates A_1 and A_2 shall define probabilites in the usual sense, it is necessary that the integral of $W(q, t)$ and of $W(p, t)$ over all values of the variables shall give a total probability of 1. Hence

$$\int |\Psi(q, t)|^2 \, dq = \int |\phi(p, t)|^2 \, dp = 1 \quad \ldots(1.12)$$

is a requirement imposed on all functions used. In equation (1.12), q represents all of the cartesian coordinates, and p represents all of the conjugate momenta. This normalization can be accomplished by application of a constant multiplier in all cases in which the integrals converge; and this convergence of the integral of the square of the absolute value of the function must be required of a function if it is to be suitable for representing the state of a mechanical system. Unless otherwise indicated, it will be assumed that all functions used are normalized in this way.

As an example, consider the function $\Psi(x) = N\exp(-\frac{1}{2}\alpha x^2)$. The constant N is called the normalization constant and must be given such a value that the square of the function is one. In this case

$$\int |\psi(x)|^2 \, dx = N^2 \int_{-\infty}^{\infty} e^{-\alpha x^2} \, dx = N^2 \left(\frac{\pi}{\alpha}\right)^{\frac{1}{2}} = 1. \quad ...(1.13)$$

From the intergral it follows that

$$\Psi(x) = \left(\frac{\alpha}{\pi}\right)^{\frac{1}{4}} e^{-\frac{1}{2}\alpha x^2} \quad ...(1.14)$$

is a properly normalized function.

It will be noticed that the function $\exp\{(i/h)\, p_\sigma x\}$, which represents a state of precisely defined momentum, does not give a convergent integral and cannot be normalized in the usual way. Neither can $\delta(x)$. These may, however, be regarded as limiting cases of functions that can be normalized. For example

$$e^{(i/h)p_0 x} = \lim_{\alpha \to 0} e^{(i/h)p_\sigma x} \, e^{-\frac{1}{2}\alpha x^2} \quad ...(1.15)$$

Functions of the Momenta

The function ϕ is as suitable as ψ for representing the state of the system, since not only is ϕ uniquely determined by Ψ, but Ψ can also be determined from ϕ. The Fourier inversion of (3.4) is

$$\psi(q_1, q_2, \ldots, q_n,)$$
$$= h^{-\frac{1}{2}n} \int dp_1 \ldots dp_n \, e^{(i/h)(p_1 q_1 + p_2 q_2 + \ldots + p_n q_n)} \phi(p_1, p_2, \ldots, p_n, t) \quad ...(1.16)$$

so that if the function ϕ were given it could be transformed to Ψ. For many applications only the functions of the coordinates are necessary.

The state represented by the function in equation (1.14) is also represented by a particularly simple function of the momentum,

$$\phi(p) = h^{-\frac{1}{2}} \left(\frac{\alpha}{\pi}\right)^{\frac{1}{4}} \int_{-\infty}^{\infty} e^{-(i/h)px} \, e^{-\frac{1}{2}\alpha x^2} \, dx$$

$$= h^{-\frac{1}{2}} \left(\frac{\alpha}{\pi}\right)^{\frac{1}{4}} e^{-p^2/2\alpha h^2} \int_{-\infty}^{\infty} e^{-\frac{1}{2}\alpha [x + (ip/\alpha h)]^2} \, dx$$

$$= \left(\frac{2}{h}\right)^{\frac{1}{2}} \left(\frac{\pi}{\alpha}\right)^{\frac{1}{4}} e^{-p^2/2\alpha h^2}$$

The function of the momentum has, in this special case, the same form as the function of the coordinates, and the spreads of the two functions are reciprocally related to each other. If α is large, the coordinate is well determined but the momentum has a very wide spread. If α is small, the situation is reversed. It will be noticed that the function ϕ in equation (1.17) is already normalized. The cofficient $h^{-\frac{1}{2}n}$ in equation (1.16) is inserted to provide that if $\Psi(x)$ is normalized, $\phi(p)$ will also be normalized.

The fact that a state can be represented by different kinds of functions is evidence of the small amount of "physical reality" to be attributed to any of them. These "wave" functions are mathematical symbols, in terms of which it is possible to represent the state of the system and to describe correctly the observed facts. The waves are not, however, simple physical things as are sound waves. One evidence of this is the fact that they are functions of all of the coordinates of the system, rather than of just three coordinates of position in real space.

Expectation Values

Instead of dealing with the probability distributions as given in Postulates A_1 and A_2 it is sometimes more convenient to deal with average values or, as they are called in statistical theory, expectation values. These are averages over all possible values, weighted by the probability of occurrence as given by the probability distribution. A knowledge of the expectation values of all the powers of a quantity is equivalent to a knowledge of the distribution function. The expectation value of any quantity whose statistical distribution is known can be obtained by multiplying each value by its probability and summing over all the values. In case a quantity, x, is distributed according to a simple error function, $P = A\exp(-\beta x^2)$, the expectation value of x is equal to

$$A\int_{-\infty}^{\infty} x \exp(-\beta x^2)dx.$$

In this case $\langle x\rangle$[1] is zero since the probability of a given positive value is just equal to the probability of the corresponding negative value. In a symmetrical distribution such as this one, the expectation value of any odd power of the variable vanishes, and the whole distribution is characterized by the averages of the even powers.

---

[1] The enclosure of a symbol such as x in brackets < > will be used to indicate a mean or expectation value.

The expectation values in a quantum-mechanical state can be determined either from the function ψ or from the function ϕ. From Postulate A_1 it follows at once that the expectation value of any power of a coordinate is given by

$$\langle q_i^s(t) \rangle = \int_{-\infty}^{\infty} \psi^*(q_1, q_2, \ldots, q_n, t)\, q_j^s\, \psi(q_1, q_2, \ldots, q_n, t)\, dq_1 \ldots dq_n \quad \ldots(1.18)$$

and from Postulate A_2 it follows at once that

$$\langle P_j^s(t) \rangle = \int_{-\infty}^{\infty} \psi^*(p_1, \ldots, p_n, t)\, p_j^s \phi(p_1, \ldots, p_n, t)\, dp_1 \ldots dp_n. \quad \ldots(1.19)$$

These expectation values are functions of the time, as is indicated. They are the averages at a definite time, when the system is in the very definite state indicated by the wave functions. These expectation values are averages over probabilities in the sense described above and are not time average over the motion.

If the expression (1.11) is used to express ϕ and ϕ^* in (1.19), there results

$$\langle p_j^s \rangle = h^{-n} \int dp \int dq' \int dq\, e^{(i/h)p \cdot q'}\, \psi^*(q') p_j^s\, e^{(-i/h)p \cdot q}\, \psi(q)$$

where dp, dq' and dq are written instead of the products of all the differentials involved, and $(p \cdot q)$ is written instead of $(p_1q_1 + p_2q_2 + p_3q_3 + \ldots p_nq_n)$. Since

$$P_j^s\, e^{-(i/h)p \cdot q} = \left(-\frac{h}{i}\right)^s \frac{\partial^s}{\partial q_j^s}\, e^{-(i/h)p \cdot q},$$

an intergration by parts leads to

$$\langle p_j^s(t) \rangle = h^{-n} \left(\frac{h}{i}\right)^s \int dp \int dq' \int dq\, e^{(i/h)(p \cdot q' - p \cdot q)}\, \psi^*(q') \frac{\partial^s}{\partial q_j^s} \psi(q).$$

This integration by parts assumes that the wave functions and their derivatives vanish for the infinites values of their arguments and that the derivatives of ψ are defined at all points. Such conditions are usually fulfilled. Integrating next with respect to p_j from $p_j = -A$ to $p_j = A$ leads to expressions of the form

$$h^{-1} \int \psi^*(q')\, dq'_j \int_{-A}^{A} dp_j\, e^{(i/h)\, p_j (q'_j - q_j)}$$

$$= \frac{2A}{h}\int \psi^*(q')\,dq'_j \frac{\sin[A(q'_j - q_j)/h]}{A(q'_j - q_j)/h}$$

In the limit as $A \to \infty$ this integrand differs from zero apprciably only in the neighbourhood of $q' = q$, so that $\psi^*(q')$ under the integral sign may be replaced by $\psi^*(q)$ and the integration over q' carried out without reference to this function. The integral is equal to 1 for each value of j; consequently

$$\langle p_j^s(t) \rangle = \left(\frac{h}{i}\right)^s \int \psi^*(q)\,\frac{\partial^s \psi(q)}{\partial q_j^s}\,dq \qquad ...(1.20)$$

Similartly it can be shown that

$$\langle q_j^s(t) \rangle = \left(\frac{h}{i}\right)^s \int \phi^*(p)\frac{\partial^s \phi(p)}{\partial p_j^s}\,dp. \qquad ...(1.21)$$

The determination of average values from the wave function can be illustrated with the function of equation. The average value of x is clearly zero and the average value of x^2 is given by

$$\langle x^2 \rangle = \left(\frac{\alpha}{\pi}\right)^{\frac{1}{2}} \int_{-\infty}^{\infty} x^2\, e^{-\alpha x^2}\,dx = \frac{1}{2\alpha}. \qquad ...(1.22)$$

The momentum averages follow directly from equation (3.16). = 0, and

$$\langle p^2 \rangle = \frac{2}{h}\left(\frac{\pi}{\alpha}\right)^{\frac{1}{2}} \int_{-\infty}^{\infty} p^2\, e^{-p^2/\alpha h^2}\,dp = \frac{1}{2}\alpha h^2. \qquad ...(1.23)$$

However, the momentum averages can also be obtained from the coordinate function $\psi(x)$ by the use of equation (3.18),

$$\langle p \rangle = \frac{h}{i}\left(\frac{\alpha}{\pi}\right)^{\frac{1}{2}} \int_{-\infty}^{\infty} e^{-\frac{1}{2}\alpha x^2}\,\frac{\partial}{\partial x}\,(e^{-\frac{1}{2}\alpha x^2})\,dx = 0$$

and

$$\langle p^2 \rangle = -h^2\left(\frac{\alpha}{\pi}\right)^{\frac{1}{2}} \int_{-\infty}^{\infty} e^{-\frac{1}{2}\alpha x^2}\,\frac{\partial^2}{\partial x^2}\,(e^{-\frac{1}{2}\alpha x^2})\,dx = \frac{1}{2}\alpha h^2 \qquad ...(1.22)$$

The above expressions for the expectation values of the momentum are also subject to the requirement that the functions of which the derivatives are taken are such that the derivatives themselves can be expressed as Fourier integrals. The function of equation (1.14) satisfies this condition since all its derivatives are continuous. However, there are many functions widely used in wave mechanics whose higher derivatives are derivatives of discontinuous functions. In such cases, the momentum distribution must be obtained from the Fourier transform.

Principle of Indetermination

The method of representing the state of a system by means of a wave function guarantees the satisfaction of the principle of indetermination because the distribution of position and the distribution of momentum are related by equation (1.14), ROBERTSON [1930], SCHROEDINGER [1930].

Consider for simplicity a case of one dimension. Let wave function be $\psi(x)$. The indetermination in x must be arbitrarily defined and the customary rms deviation may be used. Thus

$$(\Delta x)^2 = \langle (x - \langle x \rangle)^2 \rangle = \langle x^2 \rangle - \langle x \rangle^2 . \qquad \text{...(1.25)}$$

According to this definition, only the mean values of x and of x^2 are necessary to evaluate the indetermination. It is not necessary to know the distribution in any more detail. The indetermination in the momentum can be similarly defined. Let

$$(x - \langle x \rangle)\, \psi(x) = f(x),$$

and let

$$(h/\mathrm{i})\, \mathrm{d}\psi/\mathrm{d}x - \langle p \rangle \psi = g(x).$$

By the Schwartz inequality,

$$\left(\int f^* f \,\mathrm{d}x \right) \left(\int g^* g \,\mathrm{d}x \right) \geq \left| \int f^* g \,\mathrm{d}x \right|^2$$

so that

$$(\Delta x)^2 (\Delta p)^2 = \left[\int \psi^* (x - \langle x \rangle)^2 \psi \,\mathrm{d}x \right]$$

$$\times \left[\int \left(-\frac{h}{\mathrm{i}} \frac{d\psi^*}{dx} - \langle p \rangle \psi^* \right) \left(\frac{h}{\mathrm{i}} \frac{\mathrm{d}\psi}{\mathrm{d}x} - \langle p \rangle \psi \right) \mathrm{d}x \right]$$

$$\geq \left| \int \psi^* (x - \langle x \rangle) \left[\frac{h}{\mathrm{i}} \frac{\mathrm{d}\psi}{\mathrm{d}x} - \langle p \rangle \psi \right] \mathrm{d}x \right|^2$$

$$= \left| \frac{h}{\mathrm{i}} \int \psi^* x \frac{\mathrm{d}\psi}{\mathrm{d}x} \,\mathrm{d}x - \langle px \rangle \right|^2$$

$$= \left| \frac{1}{2} \frac{h}{\mathrm{i}} \left[\int \psi^* x \frac{\mathrm{d}\psi}{\mathrm{d}x} \mathrm{d}x + \int \frac{d\psi^*}{dx} x\psi \,\mathrm{d}x \right] \right.$$

$$+\frac{1}{2}\frac{h}{i}\left[\int \psi^* x \frac{d\psi}{dx}\, dx - \int \frac{d\psi^*}{dx} x\psi\, dx\right] - \langle px\rangle\Bigg|^2$$

The first bracket in the last part of this expression is real, since it is the sum of a quantity and its complex conjugate. Then, because of the imaginary factor in front of it, the whole term is imaginary. The second and third quantities are real. The square of the absolute value is equal to or greater than the square of the absolute value of the imaginary term. If the wave functions vanish in the usual way at $+\infty$ and $-\infty$ and are normalized, the first bracket can be shown by partial integration to be equal to -1. Hence for any function that is properly normalized,

$$(\Delta x)^2 (\Delta p)^2 \geq \frac{1}{4}h^2. \qquad \text{....(1.26)}$$

The exact value of the factor on the right-hand side of (3.17) depends on the arbitrary definition of Δx and Δp that has been used. The essential part is the dependence on h, which shows that the theory satisfies in this respect the requirements described in the previous chapter.

It is also of importance to notice the sign in equation (1.26). The proof given above does not show that the equality can ever be attained. However, the state represented in equation (1.6) does give this minimum value of the indetermination. From equation (1.24)

$$(\Delta p)^2 = \langle p^2\rangle - \langle p^2\rangle = \frac{1}{2}\alpha h^2$$

and from equation (1.23)

$$(\Delta x)^2 = \langle x^2\rangle - \langle x\rangle^2 = \frac{1}{2\alpha}.$$

The product of these quantities then has the minimum value of $\frac{1}{4}h^2$. Thus this state is one in which the coordinate and momentum are most accurately fixed. By suitable selection of α either x or p can be given any desired amount of precision, and the precision in the conjugate quantity will be the maximum possible.

Photoelectric Effect

The discovery of the photoelectric effect and its subsequent detailed study by many different experimenters gave evidence of still further discrepancies between the facts of observation and the implications of the electromagnetic theory of light. The experiments on the photoelectric effect showed that the energy with which an electron is ejected from a metal surface is independent of the intensity of the light and deepends on the frequency only. This is certainly not to be expected on the ordinary electromagnetic theory, since the energy density and the Poynting vector both depend upon the field strengths.

Shortly after Planck's derivation of the law of black-body radiation, Einstein [1904] proposed to treat a beam of light as a stream of corpuscles, or light quanta, each with the energy $h\nu$. He proposed to take seriously the conclusion from the observations that he energy in electromagnetic radiation occurs in bundles. Perhaps, then, these bundles have a location, a mass, and a momentum also. In the photoelectric process he proposed that a light quantum be considered as transferring all of its energy to a single electron in a process thought of as a collision. The electron will lose a certain amount, b, of this energy in escaping from the metal, so that the equation describing the process is

$$\frac{1}{2} m\upsilon^2 = h\nu - b \qquad ...(1.27)$$

where v is the velocity with which the electron escapes. This is Einstein's photoelectric law, which was verified by a series of careful measurements (MILLIKAN [1916]).

Einstein's photoelectric equation (1.27) is based on the conservation of energy only, although the model of corpuscular radiation would seem to require the conservation of momentum also. (An apparent difficulty arises here, because the momentum ascribed to a corpuscle, or photon, is $h\nu/c$. It is impossible to conserve both energy and momentum in an interaction between a single photon and a single electron if the energy of the photon is entirely transferred to the electron. This is because the energy of a photon, $h\nu$, is always just c times its momentum, while the ratio of the energy of an electron to its momentum depends on the velocity. Consequently the photoelectric process can be understood only with the assumption that the atoms of the metal take part in interaction and take up the necessary momentum along with a negligible amount of energy. Such an assumption was at first rendered dubious when Millikan showed that the connection between the quantity b and the contact potential of the metal was such as to indicate that the electrons ejected photoelectrically were the free conduction electrons. Later theoretical work reconciled these facts by showing that it is just the interaction of the conduction electrons with the surface of the metal, and hence with the forces responsible for the work function, that suffices to conserve the momentum. Hence the idea of "particles" of radiant energy is necessary and also satisfactory for the understanding of the photoelectric effect.

Compton Effect

In 1923 A. H. COMPTON [1923] showed that the scattering of X-rays could be described by the same picture as used by Einstein for the photoelectric effect. In this process the conservation of both energy and momentum must be considered, and because of high energies involved, the interactions with the atoms of the substance can be neglected except in very precise work. The process is treated as a collision between corpuscles of radiation and free electrons.

If a beam of light or X-ray is composed of corpuscles of energy $h\nu$ and momentum $h\nu/c$, these quantities will be changed in collisions with electrons. The scattered photons will have energies and momenta, and hence frequencies and wavelengths, which depend on the direction of scattering. Each electron also will recoil in a direction and with a velocity connected by the conservation laws with the direction of scattering of the photon. Let the electron be initially at rest, let the frequency of the incident photon be ν, that of the scattered photon ν', and let the velocity of recoil of the electron be v. Then the conservation of energy requires that

$$h\nu = h\nu' + mc^2 \left\{ \frac{1}{(1 - \upsilon^2/c^2)^{\frac{1}{4}}} - 1 \right\} \qquad (1.26)$$

If θ is the angle through which the photon is deflected, and if ϕ is the angle which the path of the recoil electron makes with the direction of the incident photon, the conservation of momentum requires that

$$\frac{h\nu}{c} = \frac{h\nu'}{c}\cos\theta + \frac{m\upsilon\cos\phi}{(1-\upsilon^2/c^2)^{\frac{1}{2}}}$$

$$0 = \frac{h'\nu}{c}\sin\theta - \frac{m\upsilon\sin\varphi}{(1-\upsilon^2/c^2)^{\frac{1}{2}}} \qquad ...(1.29)$$

From these relations it follows that

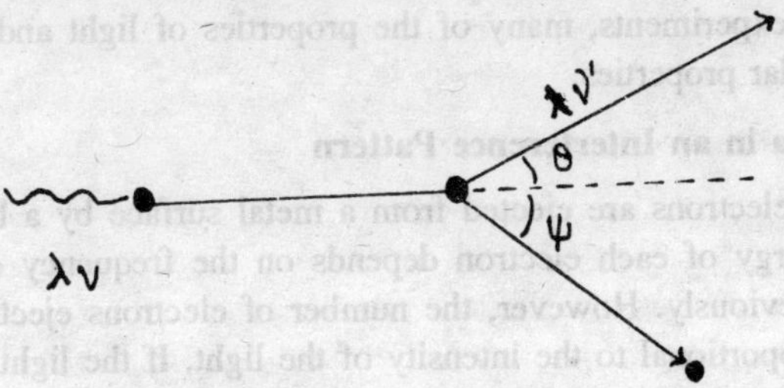

$$\frac{c(\nu - \nu')}{\nu\nu'} = \lambda' - \lambda = \frac{h}{mc}(1 - \cos\theta). \qquad ...(1.30)$$

This equation gives a relationship between the change of wavelength ($\lambda' - \lambda$) and the angle of scattering θ, which has been verified by many experiments. The angle at which the electron recoils can also be evaluated in terms of the angle θ and the frequency of the incident radiation.

Although equation (1.30) follows directly from the picture of corpuscular radiation, it is less easy to show that the experimental result could not be explained in some other way. The classical electromagnetic theory also leads to a result similar to that in equation (1.30). According to the classical treatment, the electron begins to move forward under the radiation pressure of the incident wave. The scattered wave is then changed in wavelength because of the Doppler effect. The scattered wave spreads out in all directions, and the electron recoils in the direction of the incident beam, but the relationship between the change in wavelength and the direction of the radiation is much the same as that given in equation (1.30).

To justify completely the photon picture, it was very important to establish the relationship between the direction of scattering of the radiation and the direction of recoil of the electron. This was done in 1925 by COMPTON and SIMON [1925], who actually observed by means of a Wilson cloud chamber, the tracks of the recoil electrons and the action of the scattered X-rays in producing secondary electrons.

The above three well-established but surprising experimental results, the energy distribution in black-body radiation, the photoelectric effect, and the Compton effect, produced during the first quarter of the twentieth century a conviction that in spite of the evidence of the interference experiments, many of the properties of light and X-rays are corpuscular properties.

Light quanta in an Interference Pattern

When electrons are ejected from a metal surface by a beam of light the energy of each electron depends on the frequency only, as described previously. However, the number of electrons ejected each second is proportional to the intensity of the light. If the light is very weak only one electron, on the average, may be ejected each second. It then becomes possible to count the electrons, and by inference, to count the light quanta as they strike the metal surface. With visible light the efficiency of such a counting device is low and only a small fraction of the light energy is accounted for by the ejected electrons. For shorter waves, however, such as X-rays or gamma rays, the efficiency becomes higher and light quanta can be counted.

A very impressive experiment is that in which the intensity of an interference pattern is measured by counting the light quanta incident on the screen in much the same way that alpha particles are counted.

In order to understand the distribution of light quanta over the screen one must use a wave diffraction picture in which the energy is distributed over an advancing surface. To measure the intensity, however, the number of "impacts" per second on a small area of screen is counted. This kind of edxperiment presents the dilemma of atomic physics in a striking form.

A complete description of the pheomena of electromagnetic radiation requires both a "particle"and a "wave" basis. The appropriate description depends on the particular experimental conditions. This dual description is unified by asserting that radiation of wavelength λ has quanta associated with it of momentum p, where $\lambda = h/p$.

Experiments on the Wave Properties of Particles

In spite of the impressive evidence that electrons, protons and alpha particles have the properties of small pieces of solid matter, it is possible to perform with them the same diffraction experiments as with a beam of light. If a beam of electrons is incident on a crystal, the beam is diffracted in just the manner of beam of X-rays. A narrow pencil of electrons passing through a thin film of gold to impinge on a photographic plate, produces on it, not a single spot, but a series of concentric rings just as do X-rays under similar circumstances. From the diameter of the rings, and a knowledge of the atomic spacing in the film, it is possible to compute an effective wavelength for the electron beam. The effective wavelength is equal to h/mv (DAVISSON and GERMER [1927], G.P. THOMPSON [1928]), where mv is the momentum of one of the electrons. Similar diffraction can be observed with gases, and electron diffraction is now a standard method of studying the geometricfal arrangement of atoms in molecules and solids. Perhaps even more striking is the fact that experiments can be performed with a beam of molecules, ESTERMANN and STERN [1930].

In just the same way as the photons can be counted in an optical or X-ray diffraction pattern, the electrons can be counted when they are impinging on a screen in such a way as to form a diffraction pattern. It is a relatively simple matter to count the electrons with good efficiency and such an experiment brings out the paradox of waves and particles very clearly.

Suppose the arrangement suggested in fig. 1.2 represents a source of electrons so arranged that the electrons have access to the two slits as indicated. The electrons will impinge on the screen in such a way that the number per unit time at any point is represented by the curve of intensity shown. Now suppose one of the slits is covered by a shutter. The pattern will be different, as in fig. 1.1 In some places the closing of one slit will lead to an increase in the number of electrons striking the screen; in other places, to a decrease.

There is certainly no way in which such experiments can be explained or described on the classical picture of mass points. The difficulty here is not merely one of quantitative results but is a contradiction of the very concepts of the classical theory. The contradictory nature of the situation is exemplified by the use in one equation, $\lambda = h/mv$, of quantities pertaining to waves and to corpuscles of matter. The problem of quantum mechanics is to provide a set of concepts and rules of computation that correlates these apparently contradictory phenomena into a unified whole.

Discrete Values of Energy and Magnetic Moment in Atoms

The assumption by Planck that an "oscillator" describing the electromagnetic field can have only such amplitudes of vibration as correspond to a discrete set of energy levels was the first introduction to the idea of discreteness in quantities which, in the classical theory, could take on any value. Later experiments have extended this idea in a number of ways.

Line Spectra

The striking feature of the radiation emitted by free atoms and molecules is that the spectrum consists of well defined wavelengths and frequencies of vibration. During the first part of the twentieth century, there was steady improvement in spectroscopic apparatus and a steady increase in the resolving power with which such lines could be examined. In spite of this, the spectral lines emitted by stationary and isolated atoms were shown to be very sharp indeed, and to have a "natural width" beyond the resolving power of all but the best spectrographs. These spectral lines were, at first supposed to correspond to natural frequencies of vibration of the various electrons in an atom. However, the very large number of the observed lines, and the lack of any harmonic relations between their frequencies made such a picture unsatisfactory.

About 1908, W. Ritz formulated his combination principle, according to which the frequency of a spectral line can be expressed as the difference between two spectral "terms". After Planck formulated his law for black-body radiation, it was natural to identify these spectral terms with an energy divided by *h*. Each line was then to be designated by indices representing the two terms and

$$\nu_{m,\,n} = \frac{W_m - W_n}{h} \qquad ...(1.31)$$

From this comes the idea that an atom can exist only in certain states of well defined energy, and the emission of radiation is associated with a transition between two states in which the energy difference is emitted.

Bohr's explanation of the hydrogen spectrum on this basis in 1913 was the beginning of an extensive theory of atomic structure and spectra. It proved possible to describe the hydrogen spectrum quantitatively, and other spectra qualitatively, in terms of discrete energy levels of the atoms. For atoms more complex than hydrogen, only the qualitative features of the energy-level scheme could be calculated, but the idea was sufficient for the analysis and classification of all atomic spectra.

The calculation of the discrete energy levels was made by treating classically the motion of the electrons around the heavy nucleus and then selecting, from the possible orbits, a few that were characterized by the so-called quantum rules. These rules had no rational basis, but their sucess in correlating the facts of spectroscopy made them of great importance. The general theory of quantum mechanics provides all of these results in a unified form and enable further spectral details to be worked out.

Experiments of Franck and Hertz

The experiments of FRANCK and HERTZ [1914] showed that when electrons collide with atoms they transfer to the atoms only certain discretee amounts of internal energy, which depend on the nature of the atoms. If a beam of very slow electrons is shot into a monoatomic gas, only elastic collisions can take place. As the velocity is increased, inelastic collisions begin at the definite minimum energy of excitation that such an atom can absorb. Such collisions always represent a transfer of this same amount of energy. Only when the electron energy is increased to another definite level can more energy be absorbed by the atom.

These experiments are consistent with the idea that the atoms can exist in only certain definite energy states, and the energy of these states agrees with those deduced from spectroscopy. This idea of definite energy states is quite foreign to classical mechanics. It was sometimes described as indicating a peculair stability of certain kinds of motion, but no description in terms of ordinary mechanics could be given for such stability.

Stern-Gerlach Experiment

In 1921 STERN and GERLACH [1921] sent a beam of atoms through an inhomogeneous magnetic field in such a way that an atom would be deflected by the field to an extent proportional to the component of its magnetic moment parallel to the field gradient. According to the classical conception of such a situation, the magnetic moments of the atoms would be oriented completely at random, and the beam should be drawn into a broad band whose maximum of intensity would be at the center. The experiment showed, however, a series of discrete lines corresponding to discrete values of the component of the magentic moment parallel to the field. Here again is a situation similar to that of the atomic energy levels. Only certain values of dynamical quantity seem to be permitted, while an infinity of others are for some reason excluded.

Other experiments, such as the measurement of the magnetic susceptibilities of solutions of paramagnetic salts, have confirmed the fact that the angular momentum and the magnetic moment of an atomic system can have only a discrete set of values.

Purpose of Quantum Mechanics

A careful study of experiments such as those discussed in the previous chapter makes it possible to describe conditions under which classical mechanics is applicable and conditions under which the more general quantum mechanics must be used. The precise formulation of these conditions requires the introduction of Planck's h, and in fact the classical mechanics can be understood as a limiting case of quantum mechanics, a case in which h is negligible.

Outline of Classical Mechanics[1]

Classical mechanics is based on the system of concepts and principles included in Newton's laws of motion and relativistic generalization. To apply the mechanics one must know the forces involved. In cases of natural phenomena inaccessible to experimental control, such as the motions of planets, the forces are inferred from the observed motions. In the various experiments with atomic particles the same attempt has been made to apply classical mechanics to determine the forces from observed motions. However, the observations necessary to discover the forces are less direct than in the case of planetary motion; and the results of the experiments soon raised serious doubts as to whether the classical laws are applicable to all such situations.

In classical mechanics the state of a system of mass particles is specified, at a time t, by the values of all the coordinates and all the velocities or momenta. This is, of course, an idealization.

---

[1] This brief outline of classical mechanics is inserted here only to establish a terminology and a notation to be used later, and to emphasize those portions to which quantum mechanics is closely analogous. For a more extensive discussion of these and other points, see some standard text

2

Waves and Quanta

It is possible to develop the concept of matter waves from concept of classical and quantum physics.

Matter Waves

In the previous chapter we have learnt that by 1900, the corpuscular nature of matter a had become firmly established. By 1910, Plank's quantum theory, Einsteins photoelectric equation together with various lines of experimental evidence had made it clear that light itself possesses corpuscular characteristics. By 1920 the dual nature of radiant energy was generally recognised though not understood fully.

de Broglie's Relation

In 1924, Louis de Broglie suggested that particles of matter, in particular electrons possess certain undulatory characteristics so that they too might exhibit a dual nature like the dual nature of radiant energy which was generally recognised by them. He suggested a way in which undulatory characteristics of electrons might furnish a new basis for the quantum theory.

Since nature loves symmetry so the two great entities, matter and energy, must be mutually symmetrical. Hence if energy is undulatory and/or Corpuscular matter must be Corpuscular and/or undulatory.

de Broglie developed the theory of light in terms of light quanta or light photons. To explain the phenomenon of interference he thought that light waves must be associated with photons. It was no more possible to consider that the energy is spread out over these waves, as in classical theory, but some how determine the interference pattern because the photons can produce the effect by absorbing energy. The details of interference pattern depend largely on the phase relations of these waves and de Eroglie called these as *phase waves.* He assumed there frequency V to be such that the energy in a photon is equal to $h\nu$.

de Broglie extended wave particle dualism to material particles in the only relativistically invariant way possible. He included the rest mass energy ($m_0 c^2$) in the energy, but we following Schordinger, constructed first a non-relativistic theory. We assume that any material particle of mass *m* moving at a speed υ has associated with it waves of frequency V given by

$$h\upsilon = \frac{1}{2} m\upsilon^2 + \text{P} = \text{E}$$

where *h* is the Plank's constant and P is the potential energy of the particle.

The physical nature of these waves was left indefinite by de Broglie. Perhaps they are only mathematical waves and are only used as a device for purpose of making calculation and predicting results of observations.

From the preceding discussion, the radiation may be regarded as being made up of zero-mass particles known as photons each of which carries a quantum of energy $h\nu$ and momentum $\frac{h\nu}{c} n$. It was also found that particles with nonzero mass, like the electron, have wave properties associated with them, and give diffraction patterns similar to those produced by radiation. Thus, a situation emerges wherein radiation and matter exhibit both particle properties, *i.e.* they carry quantum of energy and momentum, and wave properties, *i.e.* they produce diffraction patterns.

Now in deducing various basic laws that govern wave-particle behaviour of matter we will deduce heavily from the wave properties of radiations and particle properties of matter particles.

de Broglie's Equation

This equation can be easily derived by making use of Einsteins equation of energy ($\text{E} = mc^2$) and Plank's equation of energy ($\text{E} = h\upsilon$).

In the previous article an attempt had been made to show that some link exists between wave and particle character of a photon.

Thus we have

$$E = mc^2 \quad ...(2.1)$$

$$E = h\nu \quad ...(2.2)$$

or

$$E = \frac{hc}{\lambda} \quad ...(2.3)$$

From (1) and (3)

$$\frac{hc}{\lambda} = mc^2$$

or

$$\lambda = \frac{h}{mc} \quad ...(2.4)$$

$$= \frac{h}{p} \quad ...(2.5)$$

where p is the momentum of photon λ the wavelength is generally called *de Broglies wavelength.*

Using this relationship if calculations are made for determining the de Broglie wavelength of electron (having energy of about 1.6×10^{-10} ergs) it is found to be 1.2 Å. This distance is of the same order of magnitude as the spacing in a crystal lattice.

Making use of this relationship. Dawison and Germer showed experimentally that the electron actually possesses a wave character.

We can not answer with certainty if these waves are true waves like water waves etc.

Another Derivation of de Broglie's Equation

Let us consider a material particle, like an electron, as a standing wave system in the region of space occupied by the particle. If ψ is the quantity which undergoes periodic changes and which is responsible for matter waves. Then, the value of ψ at any instant t_0 at a point (x_0, y_0, z_0). is given by

$$\psi - \psi \sin 2\pi\, \nu_0 t_0 \quad ...(2.6)$$

where ψ_0 is the amplitude at the point chosen and ν_0 is the frequency as observed by an observer at rest with respect to the particle.

If the particle is now given a velocity v along the x-axis, then to represent the variation of ψ under new conditions, the transformation equation of relativity is taken into account

i.e.

$$t_0 = \frac{t - vx/c^2}{\sqrt{(1 - v^2/c^2)}}$$

$$\therefore \qquad \psi = \psi_0 \sin 2\pi\, \nu_0 \left\{ \frac{t - \nu x/c^2}{\sqrt{(1 - \upsilon^2/c^2)}} \right\} \qquad ...(2.7)$$

The equation of wave motion then becomes

$$y = A \sin \frac{2\pi}{T}(1 - x/\nu_p), \qquad ...(2.8)$$

Where A is the amplitude, T is the periodic time and ν_p is the velocity of the wave along the x-axis.

Comparing equations (2.7) and (2.8), we get

$$\nu_p = \frac{c^2}{\nu} \text{ and } \frac{1}{T} = \nu = \frac{\nu_0}{\sqrt{(1 - \nu^2/c^2)}} \qquad ...(2.9)$$

According to Einstein's mass energy relation, we have

$$\text{E} = h\nu_0 = m_0 c^2$$

or
$$\nu_0 = \frac{m_0 c^2}{h}$$

$$\therefore \qquad \nu = \frac{\nu_0}{\sqrt{\left(1 - \frac{\nu^2}{c^2}\right)}} = \frac{m_0 c^2}{h\sqrt{\left(1 - \frac{\nu 2}{c^2}\right)}} = \frac{mc^2}{h}$$

where
$$m = \frac{m_0}{\sqrt{\left(1 - \frac{\nu^2}{c^2}\right)}}$$

Using Eqs. (2.9) and (2.10) the wavelength of matter wave is found as follows :

$$\lambda = \frac{\text{velocity } (\nu_p)}{\text{frequency } (nu)} = \frac{c^2/\nu}{mc^2/h} = \frac{h}{m\nu} \qquad ...(2.10)$$

In the above method to obtain an expression for de Broglie wavelength it is found that a material particle in motion involves two different velocities, one of which is called the mechanical motion of the particle and is represented by ν and the other is called the propagation of the associated wave and is represented by ν_p. The two are related to each other by the following expression :

$$\nu_p = \frac{c^2}{\nu} \qquad ...(2.11)$$

Properties of de Broglie's Waves or Matter Waves

We now know that a moving particle is associated with a wave, known as **de Broglie's wave** or **matter wave.** The frequency of de Broglie's wave is given by

$$\nu = \frac{\text{Total energy}}{h} = \frac{mc^2}{h} \qquad ...(2.12)$$

The wave length of de Broglie's wave (λ) is given by

$$\lambda = \frac{h}{p} = \frac{h}{mv} \qquad ...(2.13)$$

where v is the velocity of the particle. The phase velocity of the de Broglie wave (v_p) is given by

$$v_p = \nu\lambda = \frac{c^2}{v} \qquad ...(2.14)$$

Note that the phase velocity of the wave is different from the particle velocity.

The properties of de Broglie of matter waves are given below :

(*i*) **Phase velocity.** Phase velocity

$$v_p = \frac{c^2}{v} \qquad ...(2.15)$$

and the particle velocity, $\quad v < c. \qquad ...(2.16)$

So, it follows from Eq. (2.16) that phase velocity of de Broglie waves (v_p) is always greater than the velocity of electromagnetic waves in vacuum.

(*ii*) **Dispersion in vacuum.** The phase velocity of these waves in vacuum in terms of the wavelength is given by

$$\lambda = \frac{h}{p} = \frac{h}{mv}. \quad \therefore \quad v = \frac{h}{m\lambda}$$

$$\therefore \quad v^2 \frac{h}{m_0^2 \lambda^2}\left(1 - \frac{v^2}{c^2}\right), \text{ as } m = m_0 \Big/ \left(1 - \frac{v^2}{c^2}\right)^{1/2}$$

$$\therefore \quad v^2\left(1 + \frac{h}{m_0^2 \lambda^2 c^2}\right) = \frac{h^2}{m_0{}^2 \lambda^2}$$

or $$v = \frac{h}{m_0\lambda\,(1 + h^2/m_0^2 \lambda^2 c^2)^1}$$

$$\therefore \quad \text{Phase velocity } v_p = \frac{c^2}{v} = \frac{c^2 m_0 \lambda\,(1 + h^2/m_0^2 \lambda^2 c^2)^{1/2}}{h}$$

$$= c\left(\frac{m_0{}^2 \lambda^2 c^2}{h^2} + 1\right)^{1/2}$$

Thus, we find that the phase velocity of the de Broglie waves in vacuum depends upon the wavelength, so these waves produce dispersion even in vacuum.

In the case of electromagnetic waves, the waves of all wavelengths have the same velocity in vacuum and so no dispersion of these waves occur in vacuum but dispersion occurs in a dielectric media.

(iii) **Group velocity is equal to particle velocity—Quantum picture of a material particle**

A particle while moving with a velocity v can be considered as group waves that is moving along a single direction. The group velocity of these waves is equal to the velocity of the particle.

In accordance with de Broglie hypothesis a material particle in motion is associated with a wave of wavelength $\lambda = \lambda/mv$; where m is the mass of the particle and v *velocity. If E represents the energy of the particle, then the frequency* v of the wave is given by the quantum condition $E = h\nu$,

thus $$\nu = E/h \qquad ...(2.17)$$

But according to Einstein's mass energy relation we have $E = mc^2$, so that from Eq. (2.17), we get $\nu = mc^2/h$.

Thus de Broglie wave velocity v is given by

$$v_p = \nu\lambda = \frac{mc^2}{h} \cdot \frac{h}{mv} = \frac{c^2}{v} \qquad ...(2.18)$$

According tc Einstein's theory of velocity achieved the speed of light is maximum velocity that can be by a particle in nature, thus the velocity of material particle (v) is always less than the speed of light c. Thus, from Eq. (2.18), it follows that the de Broglie wave velocity must be greater than c which is not expected. Further, it also follows that the de Broglie wave associated with the particle would travel faster than the particle itself, thereby leaving the particle far behind. Hence it is evident that a material particle will not be equivalent to a single wave train.

The above difficulty was solved by Schroedinger who postulated that a material particle in motion is equivalent to a wave packet rather than a single wave.

A wave packer is made up of a group of waves (each is having slightly different velocity and wavelength), with phases and amplitudes so chosen that they undergo interference constructively over only a small region of space where the particle can be located. Outside this space they undergo destructive interference so that the amplitude tends to zero rapidly. The amplitude of one-dimensional wave-packet, is likely to resemble the curve shown in Fig. 2.1. Such a packet moves with its own velocity v_g, known as the *group velocity*. The individual waves forming the packet possesses on average velocity, v_p known as the *phase velocity*. It can be proved that the velocity of the material particle v is same as group velocity of the group of waves.

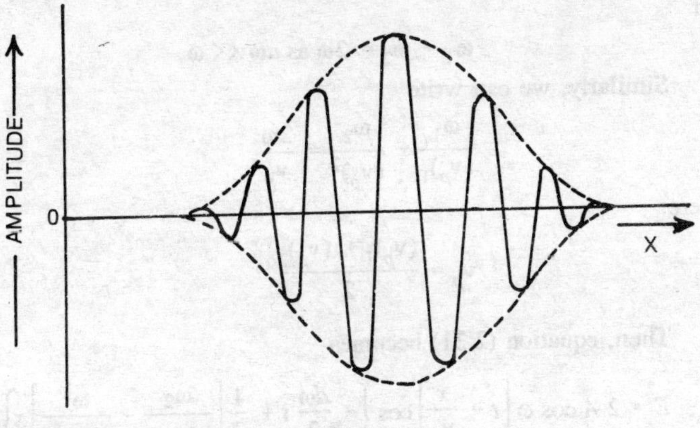

Fig. 2.1.

Consider the two wave-trains having the same amplitude; but slightly different frequencics and phase velocities. These can be represented by the equations.

$$E_1 = A \cos \omega_1 \left[t - \frac{x}{(v_p)_1} \right] \qquad ...(2.19)$$

$$E_2 = A \cos \omega_2 \left[t - \frac{x}{(v_p)^2} \right] \qquad ...(2.20)$$

The resultant amplitude of the waves can be obtained by adding the equations (1) and (2).

$$E = E_1 + E_2 = A \cos \omega_1 \left[t - \frac{x}{(v_p)_1} \right] + A \cos \omega_2 \left[t - \frac{x}{(v_p)_2} \right]$$

$$= 2A \cos\left\{\frac{\omega_1}{2}\left[t - \frac{x}{(nu_p)_1}\right] + \frac{\omega_2}{2}\left[t - \frac{x}{(nu_p)_2}\right]\right\}$$

$$\cos\left\{\frac{\omega_1}{2}\left[t - \frac{x}{(nu_p)_1}\right] - \frac{\omega_2}{2}\left[t - \frac{x}{(nu_p)_2}\right]\right\} \quad ...(2.21)$$

We can write

$$\omega_1 = \omega = 2\pi\nu$$

$$\omega_1 = \omega + d\omega = 2\pi(nu + dv)$$

Then

$$\omega_1 - \omega_2 = - d\omega$$

and

$$\omega_1 - \omega_2 = 2\omega \text{ as } d\omega << \omega$$

Similarly, we can write

$$\frac{\omega_1}{(\nu_p)_1} + \frac{\omega_2}{(\nu_p)^2} = \frac{2\omega}{\nu_p},$$

where

$$\nu_p = \frac{(\nu_p)_1 + (\nu_p)_2}{2}$$

Then, equation (2.21) becomes

$$E = 2A\cos\omega\left[t - \frac{x}{\nu_p}\right]\cos\left\{-\frac{d\omega}{2}t + \frac{1}{2}\left[\frac{\omega_2}{(nu_p)_2} - \frac{\omega^1}{(nu_p)_1}\right]x\right\}$$

Now $\quad \dfrac{\omega_2}{(\nu_p)_2} - \dfrac{\omega_1}{(\nu_p)_1} = d\left(\dfrac{\omega}{\nu_p}\right),$

so that

$$E = 2A\cos\frac{d\omega}{2}\left[t - \frac{d(\omega/\nu_p)}{d\omega}x\right]\cos\omega\left(t - \frac{x}{\nu_p}\right) \quad ...(2.22)$$

This equation represents a wave, that is travelling with phase velocity ν_p, its amplitude is given by

$$2A\cos\frac{d\omega}{2}\left[t - \frac{x}{\nu_g}\right], \quad ...(2.23)$$

where $\quad \nu_g = \dfrac{d\omega}{d(\omega/\nu_p)} = \dfrac{dv}{d(nu/\nu_p)}$

or $$\frac{f}{v_g} = \frac{d(v/v_p)}{dv}. \qquad ...(2.24)$$

Equation (2.23) shows that the amplitude itself gets propagated along X-axis with velocity v_g and since amplitude velocity is signal velocity, this is *group velocity*. Equation (6) may be written as follows:

$$\frac{1}{v_g} = \frac{d(1/\lambda)}{dv} \qquad ...(2.25)$$

$$(\lambda = v_p/v).$$

If E and V be the total energy and the potential energy of the particle respectively, then we have

$\frac{1}{2} mv^2 = E - V$, where v is the *particle velocity*

$$v = \left\{\frac{2(E - V)}{m}\right\}^{1/2} \qquad ...(2.26)$$

The de Broglie formula is

$$\lambda = \frac{h}{mv},$$

or $$\frac{1}{\lambda} = \frac{mv}{h} = \frac{m}{h}\left\{\frac{2(E - V)}{m}\right\}^{1/2}$$

[Substituting for v from Eq. (2.26)] ...(2.27)

On substituting this in equation (2.25), the group velocity is given as follows :

$$\frac{1}{v_g} = \frac{d}{dv}\left[\frac{m}{h}\left\{\frac{2(E - V)}{m}\right\}^{1/2}\right]$$

$$\frac{d}{dv}\left[\frac{m}{h}\left\{\frac{2(hv - V)}{m}\right\}^{1/2}\right] \qquad (\because \ E = hv)$$

$$\frac{1}{v_g} = \frac{1}{h}\frac{d}{dv}[\{2m(hv \times V)\}^{1/2}]$$

$$= \frac{1}{h} \cdot \frac{1}{2}\{2m(hv - V)\}^{-1/2} + 2mh$$

$$= \frac{m}{[2m(E - V)]^{1/2}}$$

$$= \left\{\frac{m}{2(E - V)}\right\}^{1/2} = \frac{1}{v}$$

$\therefore$ The group velocity v_g = The particle velocity v. ...(2.28)

From this it is clear that *a material particle in motion is equivalent to a group of waves* or a wave packet, but serious difficulty arose at once against the concept of a wave packet because a wave-packet generally gets, dissipated rapidly. For example, if a wave crest is produced at any point on a smooth surface of water, it spreads out and gets dissipated in a very short time. This if a material particle is considered as a wavepacket the existence of a *guiding wave* for it is essential because it is described by Schrodinger's equation. Without taking this assumption into consideration the wave-packet theory is not tenable. From the above it can be concluded that *the wave-packet plus guiding wave post the properties of a particle moving with velocity nu and also exhibit interference and diffraction phenomenon.* In this way the difficulty got removed.

Since the particle and wave properties are strictly complementary, so a material particle cannot behave as a particle and a wave at the same instant of time. If an experiment could be devised for showing wave properties of the matter, than its particle properties and *vice versa* could be seen. The following scheme reveals the equivalence between the wave and particle properties of the matter.

Particle Property	*Wave Property*
(*i*) momentum	wavelength
(*ii*) energy	frequency

4. Relation between phase velocity and group velocity for a non-relativistic free particle.

It v_p and v_g represent the phase velocity and group velocity for a non-relativistic free particle of mass *m*.

If λ is the de Broglie wavelength and ν the frequency of the wave. Then, the **phase velocity** is

$$v_p = \nu\lambda \quad \text{...(2.29)}$$

According to de Broglie hypothesis,

$$\lambda = \frac{h}{mv_g} \quad \text{...(2.30)}$$

Total energy E kinetic energy

$$= \frac{1}{2} m v_g^2$$

Also, $\quad E = h\nu$

So that $$\nu = \frac{E}{h} = \frac{1}{2} m v_g{}^2/h = \frac{1}{2}\frac{m v_g{}^2}{h} \quad ...(2.31)$$

Substituting of λ and ν from equations (2) and (3) in equation (1), we get

$$v_p = \frac{1}{2}\frac{m v_g^2}{h} \cdot \frac{h}{m v_g} = \frac{1}{2} v_g$$

This is the relation between phase velocity and group velocity for a non-relativistic free particle. Thus for a non-relativistic free particle the phase velocity v_p is half of the group velocity v_g.

de Broglie's Relationship and Bobr's Theory

When the de Broglie's relationship is applied to a moving electron around a nucleus it puts some restrictions on the sizes of orbits. It means that electron is not a mass particle moving in a circular path but is a standing wave train (non-energy radiating motion) extending around the nucleus in the circular path as shown in Fig. 2.2.

For the wave to be continually in phase, the circumference of the orbit should be an integral multiple of wavelength λ, *f.e.,*

$$2\pi r = n\lambda \quad ...(2.32)$$

where *r* is the radius of the orbit and *n* is a whole number.

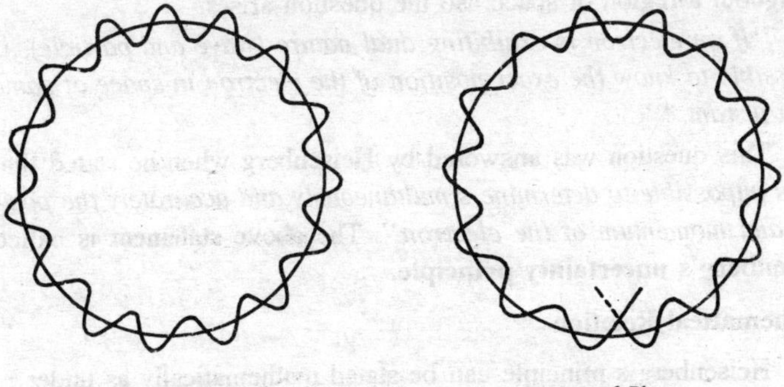

(a) In Phase (b) Out of Phase

Fig. 2.2

We know,

$$\lambda = h/mv$$

Substituting the value of λ in Eq. (2.32), we get

$$2\pi r = \frac{nh}{mv} \quad \text{or} \quad mvr = n\frac{h}{2\pi} \qquad ...(2.33)$$

which is the same as Bohr's second postulate. From equation (2.33) we find that "*Electron can move only in such orbits for which the angular momentum is an integral multiple of h/2π.*" If the circumference is bigger or smaller than the valuc given by equation (3) then the wave is not in the phase as shown in Fig. 2.2 (b). Thus, de-Broglie's relation provides a theoretical basis for the Bohr's second postulate. In Fig. 2.2 (a), the wave is in phase continually.

De Broglie Wavelength of Particles at Relativistic Velocities

The momentum of the particle will be $p = \sqrt{E^2 - E_0{}^2}/c$ where $E_0 = m_0 c^2$ is the rest energy of the particle and $E = mc^2 = E_0 + nE_k$ is the total energy of the particle.

$$\therefore \text{Wavelength,} \quad \lambda = \frac{h}{p} = \frac{ch}{\sqrt{E^2 - E_0{}^2}}$$

UNCERTAINTY PRINCIPLE

Introduction

According to the classical mechanics we can find the position and momentum of moving electron with great accuracy. However, when an electron is considered as a wave, it is, a not possible to find the exact location of the electron on the wave a is it (wave) is extending throughout a region of space. So the question arises.

"*If an electron is exhibiting dual nature (wave and particle), is it possible to know the exact position of the electron in space at same given instant ?*"

This question was answered by Heisenberg when he stated that "*It is impossible to determine simultaneously and accurately the position and momentum of the electron*". The above statement is called **Heisenberg's uncertainty principle.**

Mathematical Relation

Heisenberg's principle can be stated mathematically as under :

$$\Delta p \times x \sim \hbar \qquad ...(2.34)$$

where Δp is the uncertainty in the determination of the momentum and Δx, the uncertainty in the determination of position. Equation (2.34) is called **Heisenberg's equation** and can be stated in words as "*the product of uncertainty in the simultaneous determination of the position and momentum of a particle is equal to or greater than the $\hbar$.*" Here

$$\hbar = \frac{h}{2}\pi$$

where h is Planck's constant.

Relation (2.34) is fundamental as it sets a limit to the accurate and simultaneous measurement of position and momentum of a particle.

Heisenberg uncertainty principle made a considerable impact on our ideas of measurement. The fundamental limit ($\sim \hbar$) means not only that there is a limit to the precision of a measurement, but also that the more closely a particle gets located in position, the greater the uncertainty in the measurement of its momentum and *vice versa.*

The momentum in Equation (2.34) should actually be the component of the momentum in the x direction, and it is more properly written.

$$\Delta p_1 \times \Delta x \sim \hbar \qquad ...(2.35)$$

Pairs of variables like $p\alpha$ and x, wherein the uncertainty in one has placed a limit on the accuracy of measurement of the other, are known as *conjugate variables.* Please note that there are no limitation in the precision of measurement of the components of the momentum along the y and z axes (Δp_y and Δp_z) and the position along the x direction.

Because of the small value of h, this uncertainty has not been found to be relevant in the macroscopic world. This uncertainty is not concerned with natural uncertainty that appears in the macroscopic world, namely, the interaction that appears between the measuring instrument and the magnitude that is being measured. For example, if we try to measure the temperature of a body, we need a thermometer : but when this is kept in contact with the object, we modify the temperature we want to measure and so on.

The principle of uncertainty, which is a natural law, does not destroy the law of causality that was taken for granted in the macroscopic world. The special relativity is based solidly on the absolute distinction between cause and effect. Another name for the principle is the *indeterminacy principle.* In this respect, the classical mechanics of Newton and Galileo is *deterministic.* (If we know the force acting on a particle and the initial conditions—position and momentum—then it is possible to predict the subsequent motion of the particle with absolute precision). The microscopic world is essentially *nondeterministic.*

This problem of uncertainty is a logical consequence of the dual behaviour of matter.

Time Energy Uncertainty Principle

Just as momentum and displacement could not be found simultaneously and with infinite precision, similarly the energy E and the time t (which are another set of conjugate variables) could not be found simultaneously and with infinite precision. The more precise we are in our measurement of time (*i.e.,* the smaller Δt), the less precise we will be in the determination of energy. Again, a more sophisticated derivation would give

$$\Delta E \times \Delta t \approx \hbar \qquad ...(2.36)$$

i.e., if the time for which the system remains in a particular energy state is short, then its energy will be more defined and if it stays for longer in a state then the energy will not be so well defined. The uncertainty principle is now considered as a fundamental principle of nature. Equation (3) is applicable to complex molecules and to internal molecular energy levels as well.

For example, according to **Bohr's theory,** an electron in an atom jumps to an excited state from where it jumps back to the ground state to emit a photon of radiation. If Δt is the time for which it remains in this excited state then that will give the error in the measurement of time. In such a case the error in the measurement of energy cannot be less than $\Delta E = \frac{\hbar}{\Delta t}$. Thus, there will be always an error in the measurement of the energy of the photon, and consequently in the values of frequency and wavelength.

Even inside the nucleus, when electron gets excited, it remains in the excited state for a short duration of time Δt. Then it jumps back to the more stable state by emitting a photon of gamma ray. In estimating the energy of the photon obtained during the transition, there will always be fundamental error of the order

$$\Delta E \sim \frac{\hbar}{\Delta t}.$$

Angular momentum and angular displacement uncertainty. One other way this principle can be applied will be

$$\Delta J \times \Delta\phi \geq \hbar \quad ...(2.37)$$

where ΔJ is the error in the measurement of the angular momentum and $\Delta\phi$ is the error in the measurement of the angular displacement.

Exact Statement of Uncertainty Principle

The product of the uncertainties in determining the position and momentum of the particle can never be smaller than the number of the order $\frac{1}{2}\hbar$.

Thus, Eq. (1) may be written as

$$\Delta p \times \Delta x \geq \frac{\hbar}{2} \quad ...(2.38)$$

Similarly, Eqs. (2.36) and (2.37) may be written as

$$\Delta E \times \Delta t \geq \frac{\hbar}{2} \quad ...(2.39)$$

$$\Delta J \times \Delta\phi \geq \frac{\hbar}{2} \quad ...(2.40)$$

Proof of Heisenberg's Principle (Elementary)

We can prove the Heisenberg's uncertainty principle by assuming that a particle in motion is a group of waves and the group velocity is equal to the particle velocity.

We consider a moving body to be a de Broglie wave group and not as localised entity. Thus there is always a limit to the accuracy with which we can measure its particle properties. For example, in a de Broglie wave group in Fig. 2.3 (*a*), one cannot see center representing the position of the body very easily whereas one can measure precisely the wavelength of its component waves related to the body's momentum. However if the de Broglie wave group is very narrow as shown in Fig. 2.3 (*b*) then we can find out the position of its centre

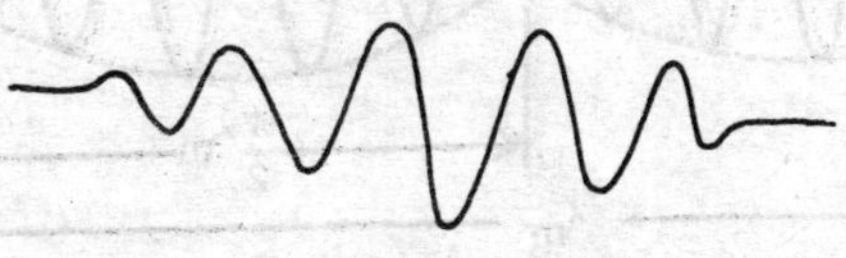

Fig. 2.3. *(a) The wavelength of a group at instant can be determined but not its position.*

readily but can never measure the wavelength.

A simple argument based upon the nature of wave groups allows

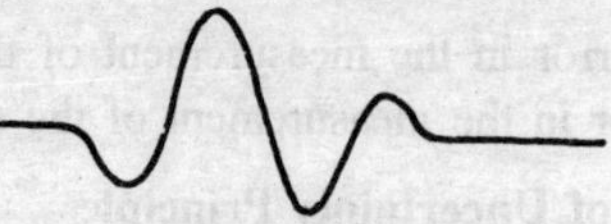

Fig. 2.3. *(b) The position of a narrow wave group at any instant can be determined but not its wavelength.*

us to relate the uncertainty Δx in a measurement of particle position with the inherent uncertainty Δp in a simultaneous measurement of its momentum. It consider two waves of angular frequencies w_1 and w_2 and propagation constants k_1 and k_2 moving along a single direction, then

$$\psi_1 = A \sin (w_1 t - k_1 x)$$

$$\psi_2 = A \sin (w_2 t - k_2 x)$$

Combining these, we get

$$\psi = \psi_1 + \psi_2 = A \sin (w_1 t - k_1 x) + A \sin (w_2 t - k_2 x)$$

$$= 2 A \sin (wt - kx) \cos \left(\frac{\delta w}{2} i - \frac{\delta k}{2} x \right)$$

where. $w = w_1 + w_2,\ k = \dfrac{k_1 + k_2}{2}$

$\delta w = w_1 \sim w_2\ \delta k = k_1 \sim k_2$.

The resultant is plotted in Fig. 2.4 where the loop formed will travel with group velocity v_g. Now this group velocity is equal to the particle velocity and so the loop formed is equivalent to the position of particle. The position of the particle cannot be given with certainty,

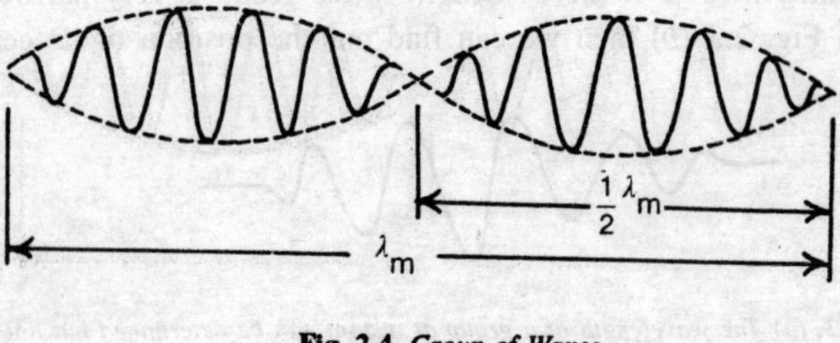

Fig. 2.4. *Group of Waxes.*

it is somewhere between one node and the next node. The error in the measurement of the position of the particle is therefore equal to the distance between these two nodes.

Now a node is formed when

$$\cos\left(\frac{\delta w}{2}\cdot t - \frac{\delta k}{2}\cdot x\right)$$

is zero. This is possible when

$$\left(\frac{\delta w}{2}\cdot t - \frac{\delta k}{2} x\right) = \frac{\pi}{2} \text{ or } \frac{3\pi}{2} \text{ or } \frac{5\pi}{2},.$$

Thus, if x_1 and x_2 represent the positions of two successive nodes, then at any instant t, we get

$$\frac{\delta w}{2}\cdot t - \frac{\delta k}{2}\cdot x_1 = (2n+1)\frac{\pi}{2}$$

and $$\frac{\delta w}{2}\cdot t - \frac{\delta k}{2}\cdot x_2 = (2n+3)\frac{\pi}{2}$$

$\therefore$ $$\frac{\delta k}{2}(x_2 - x_1) = \pi \text{ or } x_2 - x_1 = \frac{2\pi}{\delta k}$$

$\therefore$ Error in the measurement of the position of the particle is given by

$$\delta x = x_2 - x_1 = \frac{2\pi}{\delta k} = \frac{2}{2\pi\,\delta\,(1/\lambda_m)} = \frac{1}{\delta(p/h)} = \frac{h}{\delta p}$$

Thus, we obtain $\delta x.\,\delta p = h$ where δp is the error in measurement of momentum.

If we make simultaneous measurements on two quantities, *i.e.* the position and the momentum of the particle then the product of the two fundamental errors is approximately equal to h, the Planck's constant.

Note. This explanation given here is very elementary. If we consider a group of large number of waves having continuously varying frequencies, we get the product of the fundamental errors as

$$\delta p\cdot\delta x \sim \frac{\hbar}{2} = \frac{h}{4\pi}.$$

Which is used in the more accurate calculations.

SOLVED PROBLEMS

Example 1. *An electron moves in an electric field with K.E. 2.5. ev. Calculate the de Broglie wavelength associated with it.*

Solution.

$$\lambda = \frac{h}{p} = \frac{h}{\sqrt{2me \cdot E}}$$

$$= \frac{6.6 \times 10^{-27} \times 10^{8}}{\sqrt{2 \times 9.1 \times 10^{-28} + 2.5 \times 1.6 \times 10^{-12}}} \text{ Å}$$

$$= 7.734 \text{ Å}.$$

Example 2. *If an electron is to be located with in 0.1 Å, what will be the uncertainty in its velocity?*

Solution. The uncertainty in velocity in given by

$$\Delta v = \frac{\Delta p}{m}$$

$$= \frac{h}{m4x}$$

$$= \frac{10^{-27}}{10^{-27} \times 10^{-9}}$$

$$= 10^{9} \text{ cm/s}$$

Example 3. *Calculate the potential difference needed in an electron microscope to give electron a wavelength of 0.5 Å.*

Solution. K.E. of electron

$$= \frac{1}{2} m \cdot \left(\frac{h}{m\lambda}\right)^2 \qquad \left(\because \; v = \frac{h}{m\lambda}\right)$$

$$= \frac{h^2}{2m\lambda^2}$$

Here we have $\lambda = 0.5$ Å, Substitute for m and h.

$\therefore$ K.E. $= 9.66 \times 10^{-17}$ J

But K. E. $= v.\, q$

$$\therefore \quad v = \frac{\text{K E.}}{q}$$

$$= \frac{9.66 \times 10^{-17}}{5.6 \times 10^{-19}}$$

$$= 600 \text{ V}.$$

3

Wave Equations and Solutions for Some Simple Systems with Constant Potential Energy

Schrodinger's Wave Equation

We have assigned wave character to electrons so they must have the wave properties there must be a wave equation and a wave function to describe the electron waves just like the waves of light, sound and strings. Consider the motion of a string that is fixed at two ends $x = 0$ and $x = a$. It is carefully excited with certain kinds of vibrations wherein all points of the string move in such a way that their displacements change with time in the same way and all points are at their maximum displacements at the same time and possess their maximum velocity at the same time. If the displacement takes place in the y direction, mathematically these motions can be described by functions of the form

$$y(x, t) = f(x)\,\phi(t) \qquad \text{...(3.1)}$$

where $f(x)$ is independent of t and $\phi(t)$ is independent of x. Such motions are known as normal modes of vibration. The wave equation has the general form

$$\frac{d^2y}{dx^2} = \frac{T}{c^2}\frac{d^2y}{dt^2} \qquad \text{...(3.2)}$$

where c is known as the wave velocity. Substituting for y from Eq. (1) in Eq. (3.2) we get

$$\frac{c^2 d^2 f\,(x)}{f(x)\;dx^2} = \frac{1}{\phi(t)}\,\frac{d^2\,\phi(t)}{dt^2} = -\,\omega^2 \qquad \text{...(3.3)}$$

In Eq. (3.3) the variables are separated and they may be equated to the same constant, say $-\pi^2$. This gives us two ordinary differential equations

$$\frac{d^2\phi(t)}{dt^2} + \omega^2\phi(t) = 0 \qquad ...(3.4)$$

$$\frac{d^2f(x)}{dx^2} + \frac{\omega^2 f(x)}{c^2} = 0 \qquad ...(3.5)$$

This is a second order differential equation and can be solved. The solution of equation (3.4) gives

$$\phi(t) = A\sin\omega t + B\cos\omega t \qquad ...(3.6)$$

where the two constants A and B are determined from the boundary conditions, and ω is known as the circular frequency which is related to the ordinary frequency ν as under

$$\omega = 2\pi\nu \qquad ...(3.7)$$

Thus Eq. (5) may be written as

$$\frac{d^2f(x)}{dx^2} + \frac{4\pi^2\nu^2}{c^2}f(x) = 0 \qquad ...(3.8)$$

Setting $\lambda = \frac{c}{\nu}$, the general solution of Eq. (8) is

$$f(x) = A_1\exp\left(+\frac{i2\pi x}{\lambda}\right) + A_2\exp\left(-\frac{i2\pi x}{\lambda}\right) \qquad ...(3.9)$$

or
$$f(x) = C\sin\frac{2\pi}{\lambda}x + D\cos\frac{2\pi}{\lambda}x \qquad ...(3.10)$$

where A_1, A_2, C, and D are constants. It was to consider Eq. (3.10) and impose the boundary conditions.

(*i*) $f(x) = 0$ at $x = 0$; and

(*ii*) $f(x) = 0$ at $x = a$

where a is the length of the string. From the boundary condition (*i*) we get $D = 0$ and from the condition (*ii*), we get

$$C\sin\frac{2\pi}{\lambda}a = 0$$

or
$$\sin\frac{2\pi a}{\lambda} = 0$$

or
$$\frac{2a}{\lambda} = n, n = 1, 2, 3\ldots \qquad ...(3.11)$$

where n is positive integer. Thus

$$a = \frac{n\lambda}{2} \quad ...(3.12)$$

The normal modes are thus the stationary sine waves which are given by

$$f(x) = C \sin \frac{n\pi}{a} x \quad ...(3.13)$$

and the, wavelengths λ are such that length of the string is an integral number of half waves. The complete solution for a normal mode in a stretched string therefore follows from Eqs. (3.11), (3.6), (3.7) and (3.13), and is given by

$$y(x, t) = C \sin \frac{n\pi}{a} x.(A \sin 2\pi vt + B \cos 2\pi vt) \quad ...(3.14)$$

Eq. (3.14) is an equation for the amplitude of waves produced during the normal modes of vibration in a stretched string. The same equation must also represent the amplitude of a de Broglie wave associated with a moving particle. We are concerned with the time-independent or stationary waves. Therefore, the equation for a standing sine wave of wavelength λ is given by

$$\psi = C \sin \frac{2\pi}{\lambda} x \quad ...(3.15)$$

where ψ, known as the wave function, is the amplitude of the wave changing sinusiodally along x and C is the maximum amplitude. Double differentiation Eq. (15) with respect to x gives

$$\frac{a^2\psi}{dx^2} = -\frac{4\pi^2}{\lambda^2} C \sin \frac{2\pi}{\lambda} x = -\frac{4\pi^2}{\lambda^2} \psi \quad ...(3.16)$$

The kinetic energy T of a moving particle of mass m and velocity v is given by

$$T = \frac{1}{2} mv^2 = \frac{m^2v^2}{2m} \quad ...(3.17)$$

Following the de Broglie relation T becomes

$$T = \frac{h^2}{2m\lambda^2} \quad ...(3.18)$$

By using Eq. (3.16) in order to eliminate λ^2 from Eq. (18) we get

$$T = -\frac{h^2}{8\pi^2 m} \frac{1}{\psi} \frac{d^2\psi}{dx^2} \quad ...(3.19)$$

If the particle moves in a field having potential energy is v, then

$$T = E - V = -\frac{h^2}{8\pi^2 m}\frac{1}{\psi}\frac{d^2\psi}{dx^2} \qquad ...(3.20)$$

where E is the total energy. This is **Schrodinger's equation for a particle in one dimension.** It is generally written as

$$\frac{d^2\psi}{dx^2} + \frac{8\pi^2 m}{h^2}(E - V)\psi = 0 \qquad ...(3.21)$$

Schrodinger's Wave equation for a particle in three dimensions

$$\frac{d^2\psi}{dx^2} + \frac{d^2\psi}{dy^2} + \frac{d^2\psi}{dz^2} + \frac{8\pi^2 m}{h^2}(E - V)\psi = 0 \qquad ...(3.22)$$

Since $\hbar = \frac{h}{2\pi}$

making this substitution in Eq. (3.21), we get

$$\nabla^2\psi + \frac{2m}{\hbar^2}(E - V)\psi = 0 \qquad ...(3.23)$$

This is known as **Schordinger's time independent equation.** This is also called **the Schrodinger's fundamental wave equation with respect to space.**

This equation gives steady values and is found to be particularly useful when the energy of the particle is very small in comparison to the rest energy of the particle.

In case of hydrogen atom $v = -\frac{e^2}{r}$ and for it equation (3.23) becomes

$$E\psi = \left[-\frac{h^2}{8\pi^2 m}\nabla^2 - \frac{e^2}{r}\right]\psi \qquad ...(3.24)$$

when $E\psi = H\psi \qquad ...(3.25)$

$$\hat{H} = -\frac{h^2}{8\pi^2 m}\nabla^2 - \frac{e^2}{r} \qquad ...(3.26)$$

The Principle of Superposition

If we return to an equation of the form of Eq. given below

$$\nabla^2\Psi_{(xyz)} = -\frac{4\pi^2 v^2}{v^2}\Psi_{(xyz)}$$

it is readily seen that in one dimension this equation will become

$$\frac{d^2\psi_{(x)}}{dx^2} = -\frac{4\pi^2\nu^2}{v^2}\psi_{(z)}$$

This may be expressed more simpiy by setting

$$\alpha^2 = \frac{4\pi^1\nu^2}{v^2}$$

Thus we get

$$\frac{d^2\psi_{(x)}}{dx^2} + \nabla^2\psi_{(x)} = 0$$

This differential equation can be solved to get one of the solution as

$$\psi_{(x)} = A\sin\alpha x + B\cos\alpha x$$

It can be verified by carrying out the successive differentiations of the function, thus :

$$\psi_{(x)} = A\sin\alpha x + B\cos\alpha x$$

$$\frac{d\psi_{(x)}}{dx} = \alpha\,(A\cos\alpha\,x - B\sin\alpha\,x)$$

$$\frac{d^2\psi_{(x)}}{dx^2} = -\alpha^2(A\sin\alpha x + B\cos\alpha x) = -\alpha^2\psi_{(x)}$$

It can be rearranged to get

$$\frac{d^2\psi_{(x)}}{dx^2} + \alpha^2\psi_{(x)} = 0$$

Please note that

$$\psi_{(x)} = A\sin\alpha x$$

$$\psi_{(x)} = B\cos\alpha x$$

and $\psi_{(x)} = Ce^{iax}$

are other solutions of Eq. $\frac{d^2\psi_{(x)}}{dx^2}\nabla^2\psi_{(x)} = 0$. *Since according to the principle of superposition, and linear combination of solutions is also a solution, so*

$$\psi_{(x)} = A\sin\alpha x$$

and $\psi_{(x)} = Ce^{iax}$

are both solutions to the above differential equation

i.e. $$\frac{d^2\psi_{(x)}}{dx^2} + \alpha^2\psi_{(x)} = 0$$

Therefore,

$$\psi_{(x)} = A \sin \alpha x \pm Ce^{iax}$$

is also a solution to the differential equation.

This is a general solution to the differential equation as shown by the fact that the coefficients *A*, *B*, and *C* can be constants. For any equation which we are actually considering, α must also be evaluated. Thus, there is essentially an infinite number of solutions of this differential equation. However, we will see that a particular solution can be obtained by considering the boundary conditions of a given problem. In this manner specific values of the coefficients and parameters can be determined.

The Interpretation of ψ

In classical mechanics the square of wave amplitude associated with electromagnetic radiation is interpreted as a measure of the radiation intensity so we make a similar association for de Broglie waves which are associated with electrons or any particle. Let the solution of Schrodinger's wave equation in three dimensions be a function ψ (x, y, z), known as wave function. It can be anticipated that some physically observable property of the electron is connected to ψ^2 (x, y, z) or more generally ψ^* (x, y, z) ψ (x, y, z) if ψ is a complex wave function. For a system having electrons there are two ways of interpreting ψ^2 or $\psi^* \psi$. Either ψ^2 may be considered as a measure of the density of electrons of $| \psi^2 | d_\tau$ may be interpreted as a measure of probability of locating the electron in small volume d_τ in a certain region of space.

The latter view is consistent with the Heisenberg uncertainty principle and is based on results of some experiments. This principle states that the position and momentum of a small particle such as electron cannot simultaneously be measured accurately for in measuring the position of an electron we have to disturb it and so change its momentum. A complete analysis shows that if Δx and Δp_x are the uncertainties in position and momentum then $\Delta xpx \sim h$. Under these circumstances we can only predict the probability that an electron is found in a given region of space. For this reason $| \psi^2 |$ may be known as the probability function. Since electron must be somewhere in space, the integration of $| \psi^2 |$ over all space must be unity, so that

$$\int | \psi^2 | \, d_\tau = 1$$

Such wave functions are said to be normalised. For every system which is bound, every wave function must satisfy equation

$$\int |\psi^2| \, d_\tau = 1.$$

Properties of ϕ

A second order differential equation like Schrodinger's equation may give rise to several solutions though not necessarily all of them correspond to any physical or chemical reality. Such wave functions are therefore considered unacceptable. Acceptable wave functions are only those which satisfy the following conditions.

1. ψ is single valued, *i.e.*, for each value of the variables x, y, z there is only one value of the ψ. If one of the variables is an angle, say θ, then it requires that

$$\psi(\theta) = \psi(\theta + 2n\pi)$$

where n is an integer

2. ψ and its first derivative with respect to variables are continuous. In other words, there must not exist any sudden changes in ψ as its variables are changed.

3. For bound states ψ must vanish at infinity. If ψ be a complex function, then $\psi^* \psi$ must vanish at infinity.

If the above three conditions are satisfied then the function ψ is known as a well behaved function.

SOLUTIONS OF SCHRODINGER'S EQUATION ONE DIMENSIONAL PROBLEMS

Introduction

We can exactly solve schrodinger's equation for some simple systems. Such a solution is possible for such systems in which the potential energy can chosen either constant or zero. Though such systems are quite different from naturally occurring systems yet the exact solutions of simple and ideal systems give us an insight which is quite helpful in interpreting the behaviour of more complex systems which actually occur in nature.

For example the solution of Schrodinger's equation for a system of electron confined in a box having potential energy constant or zero and infinite every where else outside the box illustrates important properties of quantum mechanical systems like quantization of energy, quantum numbers, zero point energy etc.

The Particle in a One-dimensional Box

It is one of the simplest applications of wave mechanics. In it we take up the treatment of particle that is confined to move within a box. A rectangular box having dimensions *abc* along the *x, y,* and *z* axes, respectively, is chosen, and the particle is restricted to move only inside the box. So, it has no existence outside the box. Such a restriction can be met by allowing the potential energy to go to infinity at the sides of the box. This results in a *reflection* of the particle when it comes in contact with a side of the box rather than a possible *penetration.* Anywhere inside of the box the particle experiences a zero potential energy.

For simplicity, a one-dimensional box will first be considered. In the three-dimensional box, the wave function is represented by $\psi_{(xyz)}$ and in a one-dimensional box by $\psi_{(x)}$. Since the particle is to be a realistic particle like an electron, the wave function must be a function which does things a real particle will do. Such a function is called a *well-behaved* function or one belonging to class *Q.* In general, this needs that it be everywhere *continuous, finite,* and *single-valued.*

For solving a problem in wave mechanics, it is essential to solve the wave equation.

$$\nabla^2\psi_{(x\,y\,z)} + \frac{8\pi^2 m}{h^1}(E - V)\,\psi_{(x\,y\,z)} = 0 \qquad ...(3.27)$$

In case of the one-dimensional system, the wave equation becomes.

$$\frac{d^2\psi_{(x)}}{dx^2} + \frac{8\pi^2 m}{h^2}(E - V)\,\psi_{(x)} = 0 \qquad ...(3.28)$$

It is assumed that while the particle remains in the box, it has zero potential energy. Thus, the wave equation will become

$$\frac{d^2\psi_{(x)}}{dx^2} + \frac{8\pi^2 m}{h^2}E\psi_{(x)} = 0 \qquad ...(3.29)$$

This can be further simplified to give

$$\frac{d^2\psi_{(z)}}{dx^2} + \alpha^2\psi_{(z)} = 0 \qquad ...(3.30)$$

by letting $\alpha^2 = 8\pi^2 mE/h^2$. We can have an equation which is identical to Equation (3), and the solution is

$$\psi_{(x)} = A\sin\alpha x + B\cos\alpha x \qquad ...(3.31)$$

This, is a solution to the wave equation for the particle in our one-dimensional box.

So, the general solution to the differential equation provides very little information. However, we know certain restrictions which apply to this system. These are called *boundary conditions*. For instance, since the particle must not exist outside the box, it is essential for the wave function, $\psi_{(x)}$, to go to zero at the walls of the box.

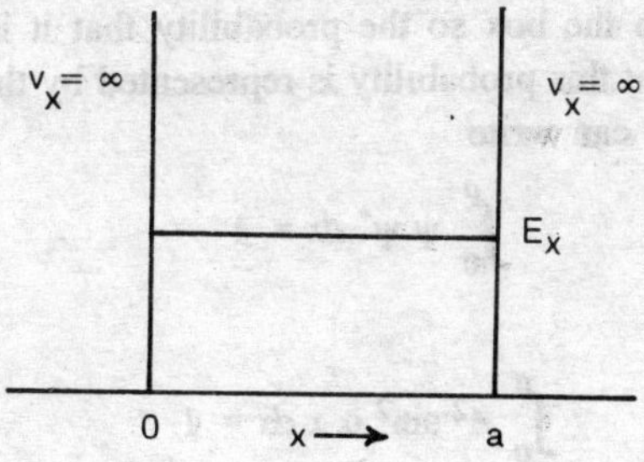

Fig. 3.1. *Particle of Energy* E_x *in a One dimensional Box.*

Thus that for the one-dimensional box shown in Figure 3.1, $\psi_{(x)} = 0$ at the point $x = 0$. Thus, we find that at the point $x = 0$,

$$0 = A \sin \alpha\, 0 + B \cos \alpha\, 0 \qquad ...(3.32)$$

$$0 = A\,(0) + B\,(1) \qquad ...(3.33)$$

For this equality to hold, it is clear that the constant, B, should be equal to zero. Thus because of this boundary condition, the wave function becomes

$$\psi_{(x)} = A \sin \alpha\, x \qquad ...(3.34)$$

At the other wall, it can be seen that the wave function must again go to zero. Thus at the point $x = a$, it is again essential that $\psi_{(x)} = 0$.

This condition offers two possible solutions. For the point $x = a$, the value function becomes

$$0 = A \sin \alpha a \qquad ...(3.35)$$

The right side of the equation can be forced to become equal zero by letting A equal zero. This would maintain the identity, but it would accomplish nothing for a useful solution. Such a solution is called a *trivial* solution. However, there is another way in which the identity can be maintained. We know that the since of an angle is zero at any integral multiple of π. Thus, if $\alpha = n\pi/a$, where n is an integer, the identity can still be satisfied. Thus by applying these boundary conditions, the wave equation for the particle reduces to

$$\psi_{(x)} = A \sin \frac{n\pi}{a} x \quad ...(3.36)$$

The only term not yet determined is the coefficient, A. It can be determined by normalizing the wave function. As it is known that the particle must be in the box so the probability that it is in the box is unity. Knowing that this probability is represented by the square of the wave function, we can write

$$\int_0^a \psi \psi^* dx = 1 \quad ...(3.37)$$

which gives

$$\int_0^a A^2 \sin^2 \alpha x \, dx = 1 \quad ...(3.38)$$

or

$$\frac{1}{A^2} = \int_0^a \sin^2 \alpha x \, dx \quad ...(3.39)$$

If this equation is solved for A and the results so obtained are substituted into the wave equation, the complete normalized wave function for the particle in a one-dimensional box is found as

$$\psi_{(x)} = \sqrt{\frac{2}{a}} \sin \frac{n\pi}{a} x \quad ...(3.40)$$

It is quite apparent that the wave function does not have to be determined so as to find the energy of the particle. This can be done if we know the value of α. The parameter can be removed from the expression $\alpha^2 = 8\pi^2 mE/h^2$ by applying $\alpha = n\pi/a$. By equating two values of α, we get

$$\frac{8\pi^2 mE}{h^2} = \frac{n^2\pi^2}{a^2} \quad ...(3.41)$$

This can be solved for the energy, to give

$$E = \frac{n^2h^2}{8ma^2} \quad ...(3.42)$$

The two significant features which must be noted with regard to the energy of the particle. Are (*i*) that the energy is quantized. Since the parameter, n, can have only integral values so the energy takes on the same type of discontinuous character which quantum theory has demanded since its rebirth with Max Planck in 1900. One of the bauties of wave mechanics is that this discontinuity results from a limited number of basic postulates and not from an *ad hoc* proposal as was essential in the Bohr model of the atom. (*ii*) a relation between the size of the box and the energy of the particle exists which may be stated as the smaller the box becomes, the greater is the energy of the particle.

Equation (3.42) gets modified if the potential energy inside the box is not zero but constant and is equal to say, v, the modified equation may be written as

$$E = \frac{n^2h^2}{8ma^2} + v \qquad ...(3.43)$$

Although the value zero for n is allowed but it is not acceptable because in that case the function ψ_n becomes, zero, but the electron is always assumed to be present inside the box. Hence the lowest kinetic energy known as *zero point energy* of an electron in a box is given by

$$E_{zero\ point} = \frac{h^2}{8ma^2} \qquad ...(3.44)$$

This equation indicates that the electron is not at rest in the box even at °K. As we only know the mean value of K. E. so the momentum of electrons is not precisely known. Thus we find that the occurrence of zero point energy is in accordance with Heisenberg's uncertainty principle.

If we remove the wall of the box and thus electron is free to move without any restriction in a field having a potential energy, zero, then the energy is given by

$$E = \frac{k^2h^2}{8\pi^2 m}$$

which is not quantised

So we conclude that when an electron is bound in a system it has quantised levels and so it gives discrete spectrum but if it has no restrictions then it has the continuous energy spectrum. This can be used to explain the occurrence of continuum in atomic and molecular spectrum. Because on ionisation an electron is free to move without any restriction.

In the above discussion, the n numbers are called *quantum numbers* while the E values are called energy levels.

If a particle is described by a wave function with a certain n value, it is said to be in the *quantum state, n.* The quantum state with lower $n(= 1)$ is known as ground state. The quantum state with higher values of $n(=2, 3,..$ etc.) are known as *excited states.*

According to quantum mechanics (but not in accordance to classical mechanics), the kinetic energy is not equal to zero in ground state. This is applicable to all problems. Moreover, if a particle is bound to a small region of space, the quantum mechanics only predicts the discrete energy levels and quantum members. On the other hand, classical mechanics does not require any such discrete energy levels for bound systems. It does not mean that quantum mechanics does not agree with classical mechanics in the region of laboratory scale sizes where classical theory is found to be applicable. In the laboratory sizes, the energy levels are so closely spaced that they could not be distinguished experimentally from a continuous set.

Given below are a few interesting facts about the foregoing problem.

(*i*) It is possible to deduce the quantum condition of discreteness of energy levels by considering the interference of de Broglie wave reflected back and forth between the walls. Some value of E not given by Eq. will give rise to distructive interference while other value of E given by Equation will give rise to constructive interference.

(*ii*) It can be compared mathematically with the vibrations of the violin string. The value of ψ represents the displacement of point on the string. As the string is fixed at each end, $y = 0$ at $x = + a$. The solution of the problem is same as that in the violin case (sinusoidal oscillations); the string is having a fundamental mode of vibration (n - 1) and set of overtones ($n = 2, 3...$) each with a characteristic frequency.

(*iii*) This can also be compared with a section of transmission line or wave guide that is shorted at both ends and here the voltage V is used in place of ψ. Further V becomes equal to zero at each end as termination becomes short (impedence $z = 0$) at each end.

Wave Function

The general form of wave equation is given below :

$$\psi_n = A \cos \frac{n\pi x}{2a} \qquad (n = 1, 3, \ldots 5) \qquad \ldots(3.45)$$

$$\psi_n = B \sin \frac{n\pi x}{2a} \qquad (n = 2, 4, \dots 6) \qquad \text{...(3.46)}$$

Let us now normalise ψ_n in Eq. (3.45)

$$\int_{-a}^{a} A^2 \cos^2 \frac{n\pi x}{2a}\, dx = 1$$

$$\int_{-a}^{a} \frac{A^2}{2}\left(1 + \cos \frac{2n\pi x}{2a}\right) dx = 1$$

$$A^2 = \frac{1}{a}$$

$$\Rightarrow \qquad A = \sqrt{\frac{1}{a}} \qquad \text{...(3.47)}$$

Similarly, normalise ψ_n in Eq. (3.46), we get

$$\Rightarrow \qquad B = \sqrt{\frac{a}{b}} \qquad \text{...(3.48)}$$

Hence normalised wave functions of Eqs. (3.45) and (3.46) are as follows :

$$\psi_n = \frac{1}{\sqrt{a}} \cos \frac{n\pi x}{2a} \qquad (n = 1, 2, 5 \dots) \qquad \text{...(3.49)}$$

$$\psi_n = \frac{1}{\sqrt{a}} \sin \frac{n\pi x}{2a} \qquad (n = 2, 4, 6 \dots) \qquad \text{...(3.50)}$$

Characteristics of Wave Functions

A few energy levels and the corresponding wave functions are shown graphically in Figure 3.2. Imagine a one dimensional mirror situated at the centre of the box and parallel to its walls (shown by dotted lines in Figure 3.2). From it we see that the wave functions are alternately symmetrical and anti-symmetrical with respect to reflections from such a mirror.

It can be seen that there are only one-half wavelength for the lowest energy state known as *zero point energy* E_0. The successive energy states differ from one another by an additional half wavelength more. There are certain points between $a \le x \le -a$ where the wave function becomes zero. These points are called *nodes*. The number of nodes, excluding those at the extremities, increases as we go up the energy scale. For a particular energy state which is characterised by the quantum number n, there will be $(n-1)$ intermediate nodes.

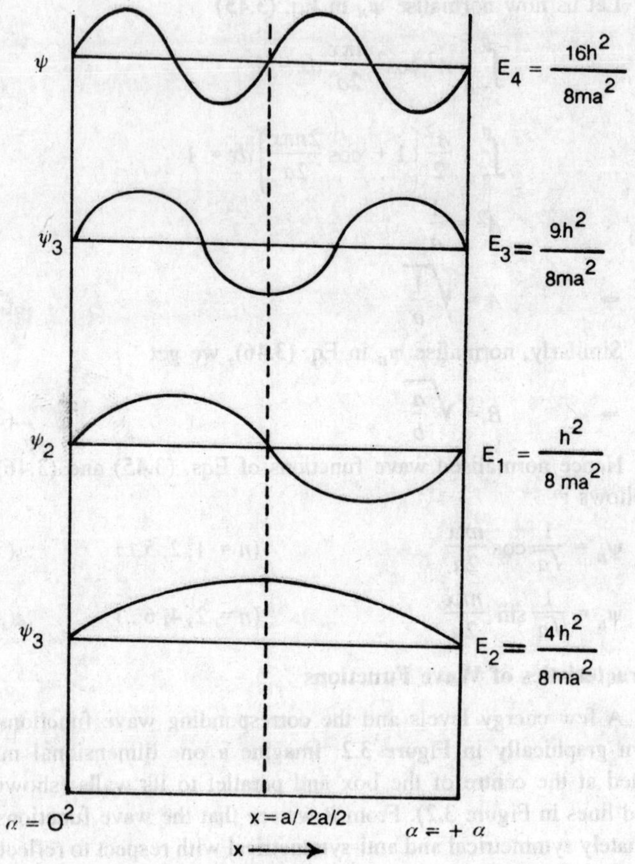

Fig. 3.2. *First few energy levels and wave functions of a particular in a box.*

As wave-functions represented by Eq. (5) are symmetrical so $\psi_n(x)$ for odd *n* are even for function of *x*. Such functions are considered to possess an *even pairty*. Similarly, $\psi_n(x)$ of odd function of *x* are anti-symmetric and are considered to be having an odd pairty.

This type of behaviour is general for all types of systems. Increasing the number of nodes decreases the wavelength which correspond to increasing the kinetic energy.

Selection Rule for an Electric Dipole Transition

We can easily show that the transition dipole integral $\int \psi_{n'}\, \bar{x}\, \psi_n\, dx$ does not vanish if $(n \pm n')$ is an odd number.

The selection rule for an electric dipole transition in such a system may be states as, "*A transition between a pair of states is allowed if the sum or difference in quantum numbers is an odd number.*"

In case both are even numbers then the transition is strictly forbidden.

The Particle in a Three-dimensional Box

For the particle in a three-dimensional box. For simplicity we consider a cubical box. The wave function in such a case will be a function of all three space coordinates. The wave equation for such a particle moving in a region of zero potential energy is given by

$$\Delta^2\, \psi_{(xyz)} + \frac{8\pi^2 m}{h^2} E\psi_{(xyz)} = 0 \qquad ...(3.51)$$

This is a partial differential equation and contains three virables. The solution of such an equation can be done in the same way as that used to separate the time and space parts of the time-dependent wave equation. Firstly we assume that the variables are separable into three individual equations, each of these containing only one variable. Each of these equations will then be a total differential equation because it will contain only one variable, and it is generally possible to find some sort of solution to these equations.

We can not find an expression always that will allow the variables to be separated, but the assumption that the total wave function can be represented as a product of wave functions is the usual place to start. For the particle in a three-dimensional box, we assumed that

$$\psi_{(xyz)} = X_{(x)}\, Y_{(y)}\, Z_{(z)} \qquad ...(3.52)$$

where $X_{(x)}$ represents a wave function which depends only on the variable, X; $Y_{(y)}$ represents a wave function that depends only on the variable, y; and so on. Substituting this expression is for $\psi_{(xyz)}$ in the wave equation we get

$$\left(\frac{\partial^2}{\partial x^2} + \frac{\partial^2}{\partial y^2} + \frac{\partial^2}{\partial z^2}\right) X_{(x)}\, Y_{(y)}\, Z_{(z)} + \frac{8\pi^2 m}{h^2} E\, X_{(x)}\, Y_{(y)}\, Z_{(z)} = 0 \qquad ...(3.53)$$

As the operator $\partial^2/\partial x^2$ has no effect on $Y_{(y)}$ and $Z_{(z)}$, and the operator $\partial^2/\partial y^2$ has no effect on $X_{(x)}$ and $Z_{(z)}$, etc., the wave equation can be rearranged to get

$$Y_{(y)}Z_{(z)}\frac{\partial^2 X_{(x)}}{\partial x^2} + X_{(x)}Z_{(z)}/\partial^2\frac{Y_{(y)}}{\partial y^2} + X_{(x)}Y_{(y)}\frac{\partial^2 Z_{(z)}}{\partial z^2} + \frac{8\pi^2 mE}{h^2}X_{(x)}Y_{(y)}Z_{(z)} = 0 \quad ...(3.54)$$

If we now divide Eq. (4) by $X_{(x)}\,Y_{(y)}\,Z_{(z)}$, we get

$$\frac{1}{E_{(x)}},\frac{\partial^2 X_{(x)}}{\partial x^2} + \frac{1}{Y_{(y)}}\frac{\partial^2 Y_{(y)}}{\partial y^2} + \frac{1}{Z_{(z)}}\frac{\partial^2 Z_{(z)}}{\partial z^2} = \frac{8\pi^2 m}{h^2}E \quad ...(3.55)$$

In Equation (3.55) each term on the left side is a function of only one variable, and the sum of these terms is the constant $-8\pi^2 mE/h^2$. So if we do not keep the variables y and z constant and allow x to vary, it can be seen that the sum of the three terms is still the same constant. Such a situation is possible only if the term $\frac{1}{X_{(x)}}\frac{\partial^2 X_{(x)}}{\partial x^2}$ is independent of x and is therefore itself a constant. The same argument will apply fully to the y and the z terms. Thus, each variable appears to be independent of the other variables, and we have suceeded in our separation. This is not a particularly unusual situation. For instance, the simple differential $\frac{d}{dx}(Ax)$ is a constant and is, therefore, independent of the variable x. If the wave function $X_{(x)}$ is substituted into the expression $\frac{1}{X_{(x)}}\frac{\partial^2 X_{(p)}}{\partial x^2}$ and the indicated differentiation done, this term can also be seen to be independent of the variable x.

Now, is the constants are represented by $-\alpha_x^2$ for the x term, $-\alpha_y^2$ for the y term, and $-\alpha_z^2$ for the z term, the following three total differential equations can be obtained.

$$\frac{1}{X_{(x)}}\frac{\partial^2 X_{(x)}}{\partial x^2} = -\alpha_x^2 \quad ...(3.56)$$

$$\frac{1}{Y_{(y)}}\frac{\partial^2 Y_{(y)}}{\partial y^2} = -\alpha_y^2 \quad ...(3.57)$$

$$\frac{1}{Z_{(z)}} \frac{d^2 Z_{(z)}}{dz^2} = -\alpha_z^2 \qquad ...(3.58)$$

From the development of our equations, it can be seen that

$$\alpha_x^2 + \alpha_y^2 + \alpha_z^2 \frac{8\pi^2 m}{h^2} E \qquad ...(3.59)$$

Thus, each degree of freedom can make its own contribution in such a way that

$$\alpha_x^2 = \frac{8\pi^2 m}{h^2} E_x \qquad ...(3.60)$$

$$\alpha_y^2 = \frac{8\pi^2 m}{h^2} E_y \qquad ...(3.61)$$

$$\alpha_z^2 = \frac{8\pi^2 m}{h^2} E_z \qquad ...(3.62)$$

Now when the variables have been separated, it becomes necessary to solve each of the equations. In this particular problem, all three of the resulting equations have the same form, so the solution of one is sufficient for demonstrating the method. Using the equation in x as an example. We rearrange it to get an equation of the exact form as the wave equation that we have solved for the one-dimensional box,

$$\frac{d^2 X_{(x)}}{dx^2} + \alpha^2 X_{(x)} = 0 \qquad ...(3.63)$$

The normalized solution then is

$$X_{(x)} = \sqrt{\frac{2}{a}} \sin \frac{nx\pi}{a} x \qquad ...(3.64)$$

and an analogous solution can be obtained for the y and z equations. Since $\psi_{(xyz)} = X_{(x)} Y_{(y)} Z_{(z)}$, the total wave function is given by

$$\psi_{(xyz)} = \sqrt{\frac{8}{abc}} \sin \frac{nx\pi}{a} x \sin \frac{n_y \pi}{b} y \sin \frac{n_z \pi}{c} Z \qquad ...(3.65)$$

It should be noted that there is a quantum number for each degree of freedom. This same idea was emphasized in the Sommerfeld quantization of the hydrogen atom, but here the quantization is a natural consequence of the mathematics.

Now that it is seen that $\alpha_x = n_x\pi/a$, $\alpha y = n_y\pi/b$ and $\alpha_x = n_z\pi/c$, the total energy for the particle in the three-dimensional box can be expressed as

$$E = E_x + E_y + E_z = \frac{h^2}{8m}\left(\frac{n_x^{\ 2}}{a^2} + \frac{n_y^{\ 2}}{b^2} + \frac{n_z^{\ 2}}{c^2}\right) \quad ...(3.66)$$

Again, we find that the energy of the particle is quantized. By choosing quantum numbers and containers of reasonable size it is found that the separation of energy levels is so small that the energy distribution will essentially be *continuous*.

In case of a cubical box $a = b = c$ and the equation (16) becomes

$$E = E_x + E_y + E_z = \frac{(nx^2 + ny^2 + nz^2)}{8ma^2} \quad ...(3.67)$$

where $n_x, n_y, n_z = 1, 2, 3, 4......$

The occurrence of three quantum numbers is characteristic of three dimensional problem in wave mechanics.

In such a case the *zero point energy* is three times that for one dimensional box because of a part associated with each of the three coordinates.

In this case a number of excited states are possible because each state is characterised by three quantum numbers. For example we can

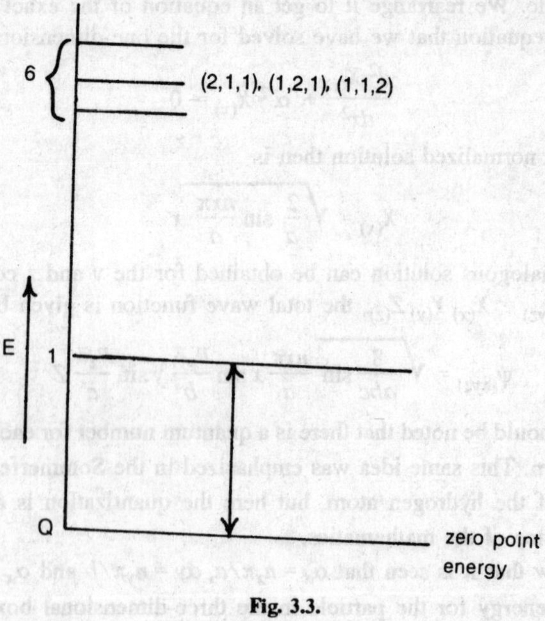

Fig. 3.3.

have three independent states characterised by quantum numbers (2, 1, 1), (1, 2, 1) and (1, 1, 2) for (n_x, n_y, n_z). The energy for all the three states is same $\left(\frac{6h^2}{8ma^2}\right)$. Each of these is three-fold degenerate (triply degenerate). The energy degeneracy and zero point energy are shown in Figure 3.3.

It has to be noted that the wave functions for the three states (triply degenerate) are different. If the cube is slightly distorted, the three fold degenerate levels get split into a non-degenerate level and doubly degenerate levels. We come across such phenomenon frequently in chemistry. Thus the electronic degeneracy is reduced or removed on slight distortion of a system.

This phenomenon is analogous to the **John-Teller effect** which states that in a non-linear molecule where electronic degeneracy occurs there always exists a vibrational mode which can remove the degeneracy.

In such a situation the molecule spontaneously distort from its most symmetric configuration until it assumes a configuration of lower symmetry as also of lower energy.

Degeneracy

We have seen that for a complete description of the energy states of a particle in a three-dimensional box, it is necessary to consider three quantum numbers. This is to be expected. The idea of quantum numbers in atomic spectra, came from our attempt to understand the positions of the spectral lines, that is, the energies represented by them. The observation of new lines led to a new quantum number that could be associated with corresponding new energy levels. Thus, we conclude that each quantum number represents a contribution to the energy of the system. However, it is generally found that for different reasons a particular set of the quantum numbers may not be unique in defining the energy of the particle. For example if the particle in a three-dimensional box is considered, the energy is given by

$$E = E_x + E_y + E_z = \frac{h^2}{8m}\left(\frac{n_x^2}{a^2} + \frac{n_y^2}{b^2} + \frac{n_z^2}{c^2}\right) \quad ...(1)$$

But for a cubical box we have $a = b = c$, and the energy can be expressed by

$$E= \frac{h^2}{8ma^2}(n_x^2 + n_y^2 + n_z^2) \quad ...(3.69)$$

For the lowest quantum state, (111), in which n_x, n_y and n_z, respectively, equal unity, it can be seen that $E= 3h^2/8ma^2$. There is only one set of quantum numbers which gives this energy state, and this level is known to be *non-degenerate*. Now consider the second energy state as shown in Figure 5.4 we can find that there are three sets (112), (121), and (211) of the quantum numbers n_x, n_y, and n_z which will give the same energy level, $E= 3h^2/4ma^2$. Such a level is known to be *degenerate*, and in this particular case, the level is *triply* degenerate. For a cubical box, it can be concluded from Figure 3.4 that virtually all of the energy levels are degenerate to some degree.

The mathematics of systems with degenerate energy states is often a problem of considerable importance. For the particle in a three-dimensional box, the degeneracy can be removed by using a box of different dimensions. If no integral relationship exists between the sides of the box, *a*, *b* and *c*, the energy levels will all be non-degenerate. Thus, it is a relatively simple problem to obtain non-degenerate energy states for the particle in a box. However, in the problems of atoms and molecules this is not always so.

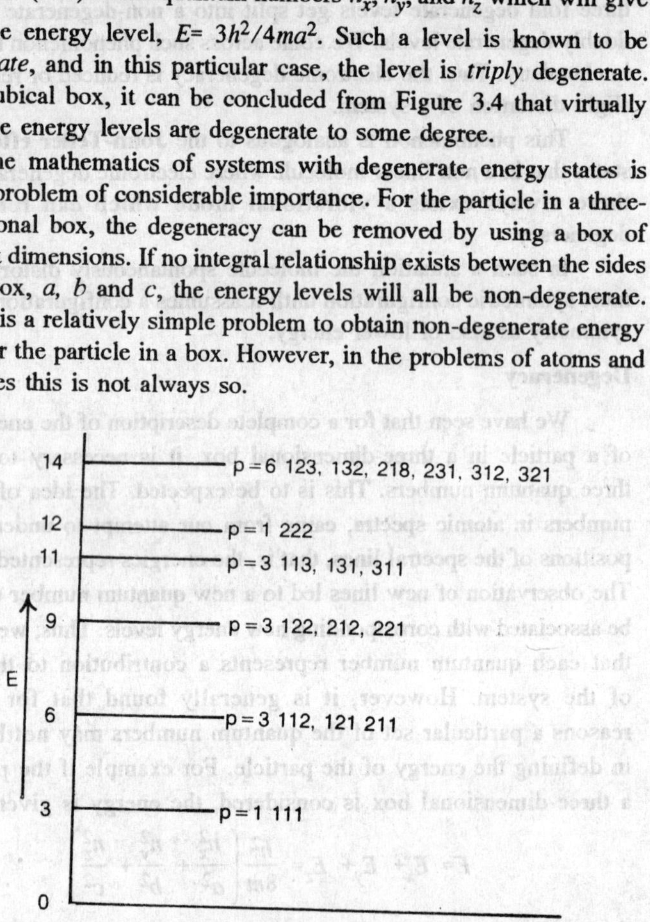

Fig. 3.4. *Degeneracy of the energy levels in a cubical box. The energy scale is in units of $h^2/8ma^2$.*

The Rigid Rotator

A rigid rotator is made up of two spherical particles attached together but are separated by finite fixed distance and capable of rotating about an axis passing through the centre of the mass and normal to the plane containing the two particles. If these two particles are forced to remain in one plane then the direction of the axis of rotation will be fixed and hence such a system is called a *rigid rotator with a fixed axis.* If the plane of these two particles can move then the axis of rotation may be free to move into any position in space and in that case the system is called *rigid rotator with free axis.*

In this section our attempt is to deal with rotation of a diatonic molecule in space. In a diatomic molecule the atoms are not fixed but vibrating with respect to each other and so the distance between atoms will not be always constant but the distance apart the equilibrium position will be constant. So in our discussion it has been assumed that the bond length remains unchanged during rotation. So the system of diatomic molecules though not actually rigid may be considered as a *rigid rotator with free axis.*

The theory of such a rigid rotator is found useful to deal with the rotational spectra of diatomic molecules.

Consider a diatomic molecule. Let the system be made up of two spherical masses, m_1 and m_2 which remain at a fixed distance apart. These two atoms of mass m_1 and m_2 are joined rigidly by a weight less link of length R. If the centre of gravity of the system is situated at the origin of the coordinate system and the two atoms of mass m_1 and m_2 are at distances r_1 and r_2 respectively from the centre of gravity, then we have

$$m_1 r_1 = m_2 r_2$$

and

$$r_1 + r_2 = \mathrm{R}$$

$$r_1 = \frac{m_2 \mathrm{R}}{m_1 + m_2}$$

$$r_2 = \frac{m_1 \mathrm{R}}{m_1 + m_2}$$

The K.E. is given by

$$\text{K. E.} = \frac{1}{2} m_1 v_1^2 + \frac{1}{2} m_2 v_2^2 \qquad \text{...(3.69a)}$$

where v_1 and v_2 are the linear velocities of the two masses m_1 and m_2 respectively.

Since we have assumed that r_1 and r_2 remain unchanged during rotation, so we can write

After making substitutions in equation (3.69a) where Θ is a function of θ only while Φ is a function of φ. If may Eq. (3.69a) is substituted in equation (3.68), and dividing both sides by $\Theta\Phi/\sin^2\theta$, then we get

$$\frac{\sin\theta}{\Theta}\frac{d}{d\theta}\left(\sin\theta\frac{d\Theta}{d\theta}\right)+\frac{2IE}{\hbar^2}\sin^2\theta$$

$$=\frac{1}{\Phi}\frac{d^2\Phi}{d\phi^2} \quad ...(3.70)$$

Let a quantity m^2 be defined by the following equation

$$\frac{1}{\Phi}\frac{d^2\Phi}{d\phi^2}=-m^2$$

i.e.,
$$\frac{d^2\Phi}{d\phi^2}+m^2\Phi=0 \quad ...(3.71)$$

and
$$\frac{\sin\theta}{\Theta}\frac{d}{d\theta}\left(\sin\theta\frac{d\Phi}{d\theta}\right)+\frac{2IE}{\hbar^2}\sin^2\theta=m^2$$

or
$$\frac{1}{\sin\theta}\frac{d}{d\theta}\left(\sin\theta\frac{d\Theta}{d\theta}\right)+\left(\frac{2IE}{\hbar^2}-\frac{m^2}{\sin^2\theta}\right)\Theta=0 \quad ...(3.72)$$

The solution of the equation (5) may be put as follows :

$$\Theta=\text{Ae}\ \bar{+}\ im\phi \quad ...(3.73)$$

when m is a positive integer.

It is possible to obtain the value of the constant A by applying the normalizing condition, *i.e.,*

$$\int\Phi\Phi^*\,d\tau=1$$

$$\therefore \quad A^2\int_0^{2\pi}e^{\,im\phi}\,e^{\,im\phi}\,d\phi=1$$

$$A^2\cdot 2\pi=1$$

$$A=\frac{1}{\sqrt{(2\pi)}} \quad ...(3.74)$$

Hence solution of equation (3.71) may be put as follows :

$$\Phi = \frac{1}{\sqrt{(2\pi)}} e^{\,im\phi}. \qquad ...(3.75)$$

Now the equation (3.72) is as follows :

$$\frac{1}{\sin\theta}\frac{d}{d\theta}\left(\sin\theta\frac{d\Theta}{d\theta}\right) + \left(\frac{2IE}{\hbar^2} - \frac{m^2}{\sin^2\theta}\right)\Theta = 0.$$

It is possible to solve this equation by making the following substitution

$$\cos\theta = x, \text{ then } -\sin\theta\, d\theta = +\, dx \text{ and}$$

$$\lambda = \frac{2IE}{\hbar^2} = l(l+1)$$

$$\therefore \quad \sin\theta\frac{d\Theta}{d\theta} = \frac{\sin^2\theta}{\sin\theta}\frac{d\Theta}{d\theta}$$

$$= -(1-x^2)\frac{d\Theta}{dx} \text{ and } \frac{1}{\sin\theta}\frac{d}{d\theta} = -\frac{d}{dx}$$

It is possible to transform the above equation into the following form

$$(1-x^2)\frac{d^2\Theta}{dx^2} - 2x\frac{d\Theta}{dx} + \left\{\lambda - \frac{m^2}{(1-x^2)}\right\}\Theta = 0 \qquad ...(3.76)$$

Eq. (3.76) is an associate legendre equation having solution as $P_l^{\,m}(x)$,

where $$P_l^{\,m}(x) = (1-x^2)^{m/2}\frac{d^m}{dx^m}P_l(x).$$

Hence the general solution of equation (6) may be put as follows:

$$\Theta_l^{\,m}(\theta)\, B\, P_l^{\,m}(\cos\theta)$$

The value of normalized constant B may be given as follows :

$$B = \left\{\frac{(2l+1)}{2}\frac{(1-|m|)!}{(1+|m|)!}\right\}^{1/2} \qquad ...(3.77)$$

$$\text{K. E.} = \frac{1}{2}m_1\omega^2 r_1^{\,2} = \frac{1}{2}m_2\omega^2 r_2^{\,2}$$

$$= \frac{1}{2}\omega^2 I$$

$$= \frac{L^2}{2I}$$

where ω is the angular velocity, I is the moment of inertial about an axis passing through the centre of gravity and normal to the line through the masses and L (*i, e.*, ω I) is the total angular momentum of rotation.

In case no force is acting on the rotator V may be taken as zero and in that case the hamiltonian may be given by

$$\hat{H} = \frac{\hat{L}^2}{2I} \quad \text{...(3.78)}$$

For solving equation (3.78) the use of spherical polar coordinates for L^2 is most convenient.

The Schrodinger's wave equation may be written as

$$\hat{H}\psi = E\psi$$

or
$$\frac{1}{\sin\theta}\frac{\partial}{\partial\theta}\left(\sin\theta\frac{\partial\psi}{\partial\theta}\right) + \frac{1}{\sin^2\theta}\frac{\partial^2\psi}{\partial\phi^2} + \frac{r\pi^2 I}{\hbar^2}E\psi = 0 \quad \text{...(3.79)}$$

Equation (2) has two variables θ and ϕ and separation of variables is carried out to solve it by assuming

$$(\psi\theta, \phi) = \Theta(\theta)\,\Phi(\theta) = \Theta\Phi \quad \text{...(3.80)}$$

Therefore,

$$\Theta_l^{\,m}(\theta) = \left\{\frac{(2l+1)}{2}\frac{(l-|m|)!}{(l+|m|)!}\right\}^{1/2} P_l^{\,m}(\cos\theta) \quad \text{...(3.81)}$$

Hence the total wave function may be put as follows :

$$\psi_{lm}(\theta, \phi)$$

$$= \left\{\frac{(2l+1)}{2}\frac{(l-|m|)!}{(l+|m|)!}\right\}^{1/2} \cdot \frac{1}{\sqrt{(2\pi)}} P_l^{\,m}(\cos\theta)\, e^{+im\phi} \quad \text{...(3.82)}$$

The corresponding eigen value of ψ_{lm} may be put as follows :

$$\frac{2IE_l}{\hbar^2} = l(1+1)$$

or
$$E_1 = \frac{\hbar^2}{2I}l(l+1) \qquad l = 0, 1, 2... \quad \text{...(3.83)}$$

when
$$l = 1 \quad E_0 = 0$$

$$l = E_1 = \frac{\hbar^2}{2I} \cdot 2 = 2B \text{ (say)}$$

$$l = 2 \quad E_2 = \frac{\hbar^2}{2I} \; 6 = 6B$$

$$I = 3 \quad E_3 = \frac{\hbar^2}{2I} \cdot 12 = 12B$$

The energy level diagram is depicted in Figure 3.5 [in which the energy levels have a characteristic ratio. Actually it is exhibited by the diatomic molecule but to rotational motion.

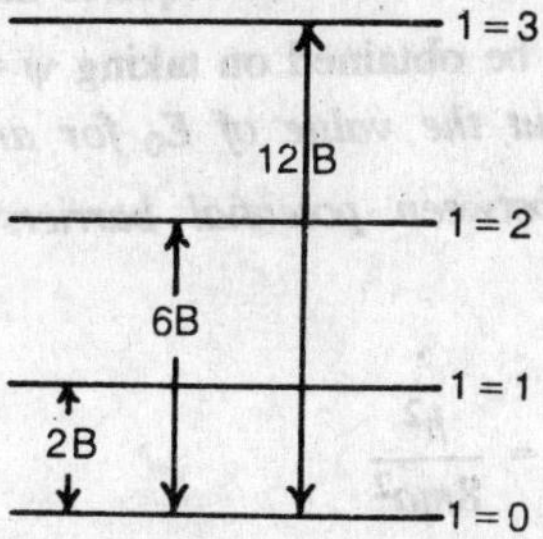

Fig. 3.5. *Energy level diagram of rigid rotator.*

SOLVED EXAMPLES

Example 1. *Consider a rigid rotator which is constrained to rotate in plane about a fixed point but is not subjected to any other forces, the energy is given by $L^2/2I$, where L is the angular momentum and I is the moment of inertia. The operator corresponding to L is $(h/2\pi i)\ (\partial/\partial\phi)$. ϕ is the angle between the rotator and the x-axis.*

(i) Determine the operator corresponding to the energy of the rotator.

(ii) Determine the allowed energy eigenfunctions of this rotator. (Consider ψ as single valued).

Solution :

(*a*) $L \rightarrow \dfrac{h}{2\pi i} \dfrac{\partial}{\partial\phi}$

$$\frac{L^2}{2I} \rightarrow - \frac{h^2}{8\pi^2 I} \frac{\partial^2}{\partial\phi^2} = H$$

(*b*) $H\psi = E\psi$

$$-\frac{h^2}{8\pi I}\frac{\partial^2\psi}{\partial\phi^2} = E\psi$$

$$\psi = \sin k\phi \text{ or } \cos k\phi \text{ or } e^{ik\phi}$$

Substituting any of these function we get

$$\frac{h^2k^2}{8\pi^2 I} = E$$

As the points $\phi = 0, \pm 2\pi, ... \pm 4\pi...$ are physically indentical, so ψ must be periodic with period 2π.

$\sin k\phi = \sin k(\phi + 2\pi) = \sin k\phi \cos 2\pi k + \cos k\phi \sin 2\pi k$.

$\cos 2\pi k = 1$ and $\sin 2\pi k = 0$. This requires that $k = 0, \pm 1, \pm 2$.

The same result can be obtained on taking $\psi = \cos k\phi$ or $e^{ik}\phi$.

Example 2. *Find out the value of E_0 for an electron that is moving back and forth between potential barriers 10^{-7} cm apart ($m = 9.1 \times 10^{-28}$ g).*

Solution :

We know $$E_0 = \frac{h^2}{8ma^2}$$

$$= \frac{(6.63 \times 10^{-27})^2}{8 \times 9.1 \times 10^{-28} \times (10^{-7})^2}$$

Ans. $= 6.04 \times 10^{-13}$ erg.

Example 3. If an electron associated with de Broglie's wavelength of 1.21×10^{-8} cm is confined in a one-dimensional box, how far apart must the walls of the box be when five loops of the de Broglie war span the distance from one wall to the other ?

Solution.

A loop is just half as long as the wavelength, *i.e.*, $1.21 \times 10^{-8}/2$ or 0.605×10^{-8} cm. If there are five loops between the walls of the box, the distance between the walls must be five times the length of the loop *i.e.*, $5 \times 0.605 \times 10^{-8}$ cm or 303×10^{-8} cm.

4

Wave Mechanics of Some Simple Systems Where the Potential Energy is not Constant

SCHRODINGERS EQUATION FOR HYDROGEN LIKE ATOMS

One Dimensional Harmonic Oscillator

We know only a few simple systems where the potential energy is not constant. The Schrodinger equation for such systems can be exactly solved. The simplest examples of such systems are the vibrations of a diatomic molecule and motions of an atom in a crystal lattice. A mode for such systems is a particle of mass *m* attracted to a point by a force *f* proportional to the displacement of the particle as

$$f = -kx \qquad ...(4.1)$$

where *k* is the force constant for the vibration and *x* is the displacement from the point of equilibrium. Eq. (4.1) is called **Hooke's law A force** of this type is known 'harmonic'. If the motion of a particle can be described by this simple law, the system is known as a 'harmonic oscillator'. The potential energy (V) of such a particle is given by the integral of this force over the distance it acts :

$$V = -\int_0^x f dx = \int_0^x kx dx = \frac{1}{2}kx^2 \qquad ...(4.2)$$

The equation of motion is

$$m\frac{d^2x}{dt^2} = -kx \qquad ...(4.3)$$

The general solution of the above equation is

$$x = a \cos 2\pi\nu_0 (t - t_0) \qquad ...(4.4)$$

where a and t_0 are constants and $\nu_0 = \frac{1}{2\pi}\sqrt{\frac{k}{m}}$, known as the fundamental frequency of the oscillator. The total energy E is given by

$$\text{E} = (1/2)\, m\left(\frac{dx}{dt}\right)^2 + 1/2\, kx^2$$

$$= 2m\pi^2{\nu_0}^2 a^2 [\sin^2 2\pi\nu_0 (t - t_0)$$

$$\cos^2 2\pi\nu_0 (t - t_0)] = 1/2\, ka^2 \qquad ...(4.5)$$

So, all positive values of E are permissible. If we examine the properties of such a system according to the laws of quantum mechanics. The hamiltonian operator for one-dimensional harmonic oscillator is given by

$$\text{H} = -\frac{h^2}{8\pi^2 m}\frac{\partial^2}{\partial x^2} + \frac{1}{2}kx^2 \qquad ...(4.6)$$

Therefore, the wave equation for this system is,

$$\frac{\partial^2\psi}{\partial x^2} + \frac{8\pi^2 m}{h^2}\left(E - \frac{1}{2}kx^2\right)\psi = 0 \qquad ...(4.7)$$

To find the well-behaved functions which satisfy Eq. (4.7) and the permissible energy levels. Eq. (4.7) can be written as given below.

$$\frac{\partial^2\psi}{\partial x^2} + (x - \beta^2 x^2)\,\psi = 0 \qquad ...(4.8)$$

where

$$x = \frac{8\pi^2 m}{h^2}E;\ \beta = \frac{2\pi}{h}\sqrt{mk} \qquad ...(4.9)$$

By changing variables $\xi = \sqrt{\beta x}$ we can rewrite Eq. (4.8) as

$$\frac{\partial^2\psi}{\partial \xi^2} + \left(\frac{\alpha}{\beta} - \xi^2\right)\psi = 0 \qquad ...(4.10)$$

Eq. (4.10), can be solved as follows. Consider a function :

$$\lambda = \exp\,(-x^{12}) \qquad ...(4.11)$$

Differentiating we get

$$\frac{\partial y}{\partial x'} + 2x'\, y = 0 \qquad ...(4.12)$$

Differentiating $(n + 1)$ more times, we get

$$\frac{\partial^2 Z}{\partial x'^2} + 2x' \frac{\partial Z}{\partial x'} + 2(n+1) Z = 0 \qquad ...(4.13)$$

where $$Z = \frac{\partial m}{\partial x} = \frac{\partial n}{\partial x'n} [\exp(-x^{12})] \qquad ...(4.14)$$

Assume a function $\frac{\partial ny}{\partial x'n}$

$$U = (-1)^n \exp\left(\frac{x'^2}{2}\right) \cdot Z \qquad ...(4.15)$$

Thus we have

$$\frac{\partial^2 U}{\partial x'^2} = \exp\left(\frac{x'^2}{2}\right)\left[\frac{\partial^2 Z}{\partial x'^2} + 2x' \frac{\partial Z}{\partial x'}\right]$$

$$+ Z. \exp\left(\frac{x'^2}{2}\right)(x'^2 + 1) \qquad ...(4.16)$$

or $$\left[\frac{\partial^2 Z}{\partial x'^2} + 2x' \frac{\partial Z}{\partial x'}\right]$$

$$= \exp\left(-\frac{x'^2}{2}\right)\frac{\partial^2 U}{\partial x'^2} - Z(x'^2 + 1) \qquad ...(4.17)$$

Inserting Eq. (4.17) into (4.13) we get

$$\frac{\partial^2 U}{\partial x'^2} + [2n + 1 - x'^2]\, U = 0 \qquad ...(4.18)$$

Comparing Eq. (18) with Eq. (10), we find that the two equations are the same (except for the variables), Hence

$$\frac{\alpha}{\beta} = 2n + 1 \qquad ...(4.19)$$

or $$E = \frac{h}{2\pi}\sqrt{\frac{k}{m}} \cdot \left(n + \frac{1}{2}\right) \qquad ...(4.20)$$

Since the classical frequency v_0 of a vibration is

$$\frac{1}{2\pi}\sqrt{\frac{k}{m}},$$

we can write

$$E = \left(n + \frac{1}{2}\right) hv_0 \qquad ...(4.21)$$

where $n = 0, 1, 2, 3 \ldots$ and is known as the vibrational quantum number. The state with $n = 0$ is the vibrational ground state and it possesses energy $(1/2)\, h\nu_0$, known as the *zero-point energy*. The occurrence of zero-point energy is consistent with uncertainty principle.

The solution ψ of Eq. (4.10) is of the form of the function U, given by Eq. (4.15). We have

$$\psi = (-1)^n \exp(\xi^2/2) \frac{\partial^n}{\partial \xi^n} [\exp(-\xi^2)]$$

$$= (-1)^n \exp(-\xi^2/2) \exp(\xi^2)$$

$$\frac{\partial^n}{\partial \xi^2} [\exp(-\xi^2)] \qquad \ldots(4.22)$$

The Hermite polynomial of degree n can be defined as

$$H_n(\xi) = (-1)^n \exp(\xi^2) \frac{\partial^n}{\partial \xi^v} [\exp(-\xi^2)] \qquad \ldots(4.23)$$

Some Hermite polynomials are :

$$\left.\begin{array}{ll} H_0(\xi) = 1 & H_2(\xi) = 4\xi^2 - 2 \\ H_1(\xi) = 2\xi & H_3 8(\xi) = \xi^3 - 12\xi \\ H_4(\xi) = 16\xi^4 - 48\xi^2 + 12 & \end{array}\right\} \qquad \ldots(4.24)$$

Therefore, the *un-normalised wave function of a harmonic oscillator* in its nth quantum state of vibration can be written as

$$\psi_n(\xi) = \exp(-\xi^2/2)\, H_n(\xi) \qquad \ldots(4.25)$$

when the first factor $\exp(-\xi^2/2)$ possesses no nodes but it forces ψ to approach zero as ξ tends to $\pm\,\alpha$. On the other hand the function $H_n(\xi)$ introduces n nodes into ψ_n because it is a polynomial with n real roots. Because n can have any integral value or zero, our solution gives the required number of nodes.

The solution of three dimensional oscillator does not create any new problem if the coupling between the oscillations in the directions of the basis vectors, x, y and z is not taken into consideration. Such an oscillator is known as an isotropic oscillator. The treatment of an enharmonic oscillator wherein the driving force does not follow Hooke's law cannot be made because the wave equation in such a case does not have the exact solution. It is, therefore, necessary to use the approximate methods for solving the wave equation for the anharmonic oscillator.

Normalisation and the Characteristics of the Eigenfunctions of a Harmonic Oscillator

We know that normalisation condition of the wave functions for the oscillator is

$$\int_{-\infty}^{\infty} \psi_n^*(\xi)\psi_n(\xi)d\xi = 1 \qquad ...(4.26)$$

un-normalised wave function of harmonic oscillator is given by

$$\int_{-\infty}^{\infty} \exp(-\xi^2)\,[H_n(\xi)]^2\, d\xi = 2^n n!\sqrt{\pi} \qquad ...(4.27)$$

So the normalised wave functions for the harmonic oscillator are

$$\psi_n(\xi) = \left[\frac{1}{2^n n!\sqrt{\pi}}\right]^{1/2} \exp(-\xi^2/2)H_n(\xi) \qquad ...(4.28)$$

in the range $-\infty < \xi < +\infty$. It can be shown that for $n \neq n'$, the wave functions are orthogonal, *i.e.*

$$\int_{-\infty}^{\infty} \psi_n^*(\xi)\psi_n'(\xi)\, d\xi = 0 \qquad ...(4.29)$$

The validity of Eq. (4) is justified because the functions ψ_n and ψ_n' are the eigenfunctions of a hamiltonian which is hermitian and posses-ses different eigenvalues.

The first few energy levels and the corresponding wave functions are shown in Fig. 4.1.

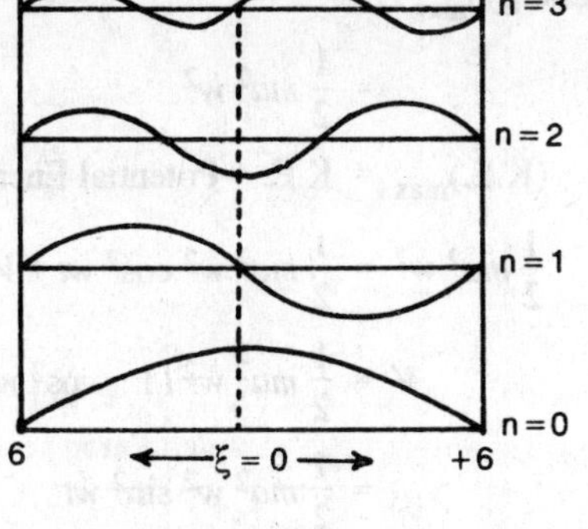

Fig. 4.1. *Energy levels and wave functions of the harmonic oscillator.*

It can be seen that the wave functions characterised by an add vibrational quantum number are antisymmetrical and those characterised by either zero or an even quantum number are symmetrical about the origin ($\xi = 0$).

ASYMPTOTIC BEHAVIOUR OF A LINEAR HARMONIC OSCILLATOR

Asymptotic Behaviour

For a linear harmonic oscillator, we can represent the force $F = -Kx$ by the potential energy

$$V(x) = \frac{1}{2} Kx^2 \quad ...(1)$$

Above equation can be derived as under.

Let us consider a particle of mass m which is undergoing simple harmonic motion with amplitude a. Its displacement from the origin at time t is given by

$$x = a \sin wt$$

or

$$\frac{dx}{dt} = aw \cos wt$$

$$\therefore \text{ Kinetic energy } = \frac{1}{2} m \left(\frac{dx}{dt}\right)^2$$

$$= \frac{1}{2} m a^2 w^2 \cos^2 wt$$

Maximum value of K.E. will be for

$$\cos wt = 1$$

$\therefore$ (Kinetic Energy)$_{max}$

$$= \frac{1}{2} ma^2 w^2$$

Moreover, $\quad (\text{K.E.})_{max} = \text{K.E.} + \text{Potential Energy } (V)$

or

$$\frac{1}{2} ma^2 w^2 = \frac{1}{2} ma^2 w^2 \cos^2 wt + V$$

or

$$V = \frac{1}{2} ma^2 w^2 (1 - \cos^2 wt))$$

$$= \frac{1}{2} ma^2 w^2 \sin^2 wt$$

$$= \frac{1}{2} mw^2x^2 \qquad [\because \; x = \sin wt]$$

$$= \frac{1}{2} Kx^2 \qquad ...(4.31)$$

where $K = mw^2$. It is to be proved.

If we substitute Eq. (4.31) in the Schroedinger's equation in one dimension, we get

$$\frac{d^2\psi}{dx^2} + \frac{2m}{\hbar^2}\left(E - \frac{1}{2} Kx^2\right)\psi = 0 \qquad ...(4.32)$$

Eq. (4.32) can be changed in dimensionless form by introducing

$$\xi = \alpha x$$

$$\therefore \quad \frac{d\psi}{dx} = \frac{d\psi}{d\xi}\frac{d\xi}{dx} = \frac{d\psi}{d\xi}\alpha$$

and
$$\frac{d^2\psi}{dx^2} = \frac{d}{dx}\left(\frac{d\psi}{a\xi}\alpha\right)$$

$$= \frac{d}{d\xi}\frac{d\xi}{dx}\left(\frac{d\psi}{dx}\alpha\right) = \alpha^2\frac{d^2\psi}{d\xi} \qquad ...(4.33)$$

Introducing these in Eq. (4.32), we get

$$\alpha^2\frac{d^2\psi}{d\xi^2} + \left[\frac{2mE}{\hbar^2} - \frac{mK}{\hbar^2}\frac{\xi^2}{\alpha^2}\right]\psi = 0$$

or
$$\frac{d^2\psi}{d\zeta^2} + \left[\frac{2mE}{\hbar^2\alpha^2} - \frac{hK\xi^2}{\alpha_0^2\hbar^4}\right]\psi = 0 \qquad ...(4.34)$$

If α is so chosen that

$$\frac{mK}{\hbar^2\alpha^4} = 1 \text{ or } \alpha^4 = \frac{mK}{\hbar^2}, \text{ i.e., } \alpha = \left[\frac{mK}{\hbar^2}\right]^{1/4}$$

and if
$$\lambda = \frac{2mE}{\hbar^2\alpha^2}$$

or
$$\lambda = \frac{2mEh}{\hbar^2\sqrt{(mK)}} = \frac{2E}{K}\left(\frac{m}{K}\right)^{1/2} \qquad ...(4.35)$$

If Eq. (4.35) is substituted in Eq. (4.34), we get

$$\frac{d^2\psi}{d\xi^2} + (\lambda - \xi^2)\,\psi = 0. \qquad ...(4.36)$$

Now we aim to find such functions which can satisfy this equation throughout the region of value $-\infty$ to $+\infty$ for ξ and are acceptable wave functions. Thus we desire such functions which are continuous, single valued and finite throughout the region.

Now a solution of Eq. (4.36) is attempted which is valid through out configuration space $(-\infty < \xi < +\infty)$. This is based upon the asymptotic behaviour. Suppose the solution is of the following form :

$$\psi(\xi) = H(\xi)e^{-\xi^2/2}. \qquad ...(4.37)$$

where $H(\xi)$ stands for a polynomial of finite order in ξ. If positive sign is taken in the exponent, ψ get diverged as $\xi \to \infty$.

From Eq. (4.34), $\dfrac{d\psi}{d\xi} = H'(\xi)\ \text{exp.}\left[\dfrac{-\xi^2}{2}\right] - \xi H(\xi)\ \text{exp.}\left[\dfrac{-\xi^2}{2}\right]$

and $$\frac{d^2\psi}{d\xi^2} = H'(\xi)\ \text{exp.}\left[\frac{-\xi^2}{2}\right] - H'(\xi)\xi\ \text{exp.}\left[\frac{-\xi^2}{2}\right]$$

$$- H(\xi)\ \text{exp.}\left[\frac{-\xi^2}{2}\right] - \xi H'(\xi)\ \text{exp.}\left[\frac{-\xi^2}{2}\right] - \xi H(\xi)\ \text{exp.}\left[\frac{-\xi^2}{2}\right]$$

$$= H''(\xi)\ \text{exp.}\left[\frac{-\xi^2}{2}\right] - 2\xi H'(\xi).\ \text{exp.}\left[\frac{-\xi sup}{2}\right]$$

$$= H''(\xi)\ \text{exp.}\left[\frac{-\xi^2}{2}\right] - 2\xi H'(\xi)\ \text{exp.}\left[\frac{-\xi sup}{2}\right]$$

$$+ (\xi^2 - 1)\,H(\xi)\ \text{exp.}\left[\frac{-\xi^2}{2}\right].$$

If the values of $\dfrac{d\psi}{d\xi}$ and $\left[\dfrac{d^2\psi}{d\xi^2}\right]$ are substituted in Eq. (4.36), we get

$$\text{exp.}\left[\frac{-\xi^2}{2}\right]\left[H''(\xi) - 2\xi H'(\xi) + \xi^2 H(\xi) - H(\xi) + (\lambda - \xi^2)\,H(\xi)\right] = 0$$

or $$H''(\xi) - 2\xi H'(\xi) + (\lambda - 1)H(\xi) = 0$$

or $$H'' - 2\xi H' + (\lambda - 1)\,H = 0, \qquad ...(4.38)$$

In Eq. (4.38) primes represent differentiation with respect to ξ.

Energy levels. Eq. (4.35) can also be solved by the series method by putting

$$H(\xi) = \xi^s (a_0 + a_1\xi + a_2\xi^2 + ...) \qquad a_0 \neq 0, s \geq 0$$

$$= \xi^s \sum_{\nu=0}^{\infty} a_\nu \xi^\nu$$

$$= \sum_{\nu=0}^{\infty} a_\nu \xi^{s+\nu} \qquad ...(4.39)$$

$$\frac{dH}{d\xi} = \sum_\nu a_\nu (s+\nu)\, \xi^{s+y-1}$$

and
$$\frac{a^2H}{d\xi^2} = \sum_\nu a_\nu (s+\nu)(s+\nu-1)\, \xi^{s+\nu-2}.$$

Substituting these values in equation (9), we get

$$\sum_\nu a_\nu (s+\nu)(s+\nu-1)\, \xi^{s+\nu-2} - 2\xi \sum_\nu a_\nu (s+\nu)\, \xi^{s+\nu-1}$$
$$+ (\lambda - 1) \sum_\nu a_\nu \xi^{s+\nu} = 0$$

or
$$\sum_\nu a_\nu (s+\nu)(s+\nu-1)\, \xi^{s+\nu-2} - 2 \sum_\nu a_\nu (s+\nu)\, \xi^{s+\nu-1}$$
$$+ (\lambda - 1) \sum a_\nu \xi^{s+\nu} = 0 \qquad ...(4.40)$$

For this series to vanish for all values of ξ *i. e.*, $H(\xi)$ must be a solution of Eq. (4.38), it means that the coefficient of individual powers of ξ must vanish separately. If the coefficient of different powers of ξ are equated to zero, we get

$$s(s-1)\, a_0 = 0 \qquad ...(4.41a)$$

$$(s+1)(s)\, a_1 = 0 \qquad ...(4.41b)$$

$$(s+2)(s+1)\, a_2 - (2s+1+\lambda)\, a_0 = 0 \qquad ...(4.41c)$$

$$(s+3)(s+2)\, a_2 - (2s-3-\lambda)\, a_1 = 0 \qquad ...(4.41d)$$

$$\cdots \qquad \cdots \qquad \cdots \qquad \cdots$$

$$\cdots \qquad \cdots \qquad \cdots \qquad \cdots$$

$$(s+\nu+2)(s+\nu+1)\, a_{\nu+2}$$
$$- (2s+2\nu+1-\lambda)\, a_\nu = 0 \qquad ...(4.41e)$$

$$a_{u+2} = \frac{2s+2\nu+1-\lambda}{(s+\nu+1)(s+\nu+2)}\, a_\nu \qquad ...(4.42)$$

where ν is an integer.

Eq. (4.42) is called a *recursion formula.* Since a_0 cannot be zero, so it follows from Eq. 4.41(*a*) that $s = 0$ or $s = 1$. If a_0 is put equal to zero, only odd powers will appear; with *a*, zero, the series will be having even powers only.

If arbitrary values are given to the energy parameter λ, the series given in Eq. (4.42) will give rise to an infinite number of terms but they do not correspond to satisfactory wave function. If the convergence of the power solution defined by Eq. (4.42) is examined carefully, it is found that as $v \to \infty$, $(a_{v+2})/a_v \to 2/v$ so that the series gets converged for all finite values of ξ. Now we will consider the series expansion of $e^{+\xi^2}$, *i.e.*,

$$e^{\xi^2} = 1 + \xi^2 + \frac{\xi^4}{2\,!} + \frac{\xi^2}{3\,!} + \ldots + \frac{\xi^v}{(v/2)\,!} + \frac{\xi^{v+2}}{(v/2+1)\,!} + \ldots$$

$$= b_0 + b_2\,\xi^2 + b_4\,\xi^4 + \ldots + b_v\,\xi^v + b_v\,\xi^v + b_{v+2}\,\xi^{v+2} + \ldots$$

and
$$\frac{b_{v+2}}{bv} = \frac{\dfrac{1}{(v/1+1)\,!}}{\dfrac{1}{(v/2)\,!}} = \frac{2}{2+v}$$

or
$$\lim_{v \to \infty} \frac{b_{v+2}}{b_v} = \frac{2}{v}.$$

From the above it follows that $H(\xi)$ gets diverged approximately as e^{ξ^2} and the product $H\ (\xi)e^{-\xi^2/2}$ will behave in a similar manner as $e^{+\xi^2/2}$ in this region. It means that it is not acceptable as wave function.

To avoid this situation, the value of λ is so chosen that the power series for $H(\xi)$ gets cut off at some terms, thereby making $H(\xi)$ a polynomial.

From Eq. (4.42), it follows that the value of λ which makes the series to cut off at *n*th term, should be

$$\lambda = 2s + 2v + 1$$

The value of index, *s*, may be either 0 or 1. Corresponding to these values of *s* the values of λ will be equal to $2v + 1$ or $2v + 3$ where $2v$ is an even integer. It is possible to express both these cases in terms of a quantum number *n*, *i.e.*,

$$\lambda = (2n + 1).$$

Then
$$\frac{2E_n}{\hbar}\sqrt{\frac{m}{K}} = (2n + 1)$$

or $$E_n = (n + \frac{1}{2})\hbar\omega \qquad n = 0, 1, 2, 3, \qquad ...(4.43)$$

where $$\omega_c = \sqrt{\frac{K}{m}}$$

is the classical angular frequency of the oscillator.

Zero Point Energy

All the energy levels obtained by using Eq. (4.43) are discrete and are equally spaced. If $n = 0$ is put in Eq. (4.43), then the finite value of ground state energy will be $\frac{1}{2}\hbar\omega_0$, which is known as the *zero point energy i.e.,*

$$E_0 = \frac{1}{2}\hbar\omega_0 \qquad ....(4.44)$$

From this equation it is clear that all tne energy levels get shifted by an amount which is equal to half the separation of energy levels.

In Fig. 6.2 five lowest energy levels of a harmonic oscillator are shown. From this figure, it can be usualised that the system in its lowest state has an energy greater than that which it would be having when it was at rest in its equilibrium position. Zero point energy has been found to be characteristic of quantum mechanics Moreover it is related to the Heisenberg's uncertainty principle.

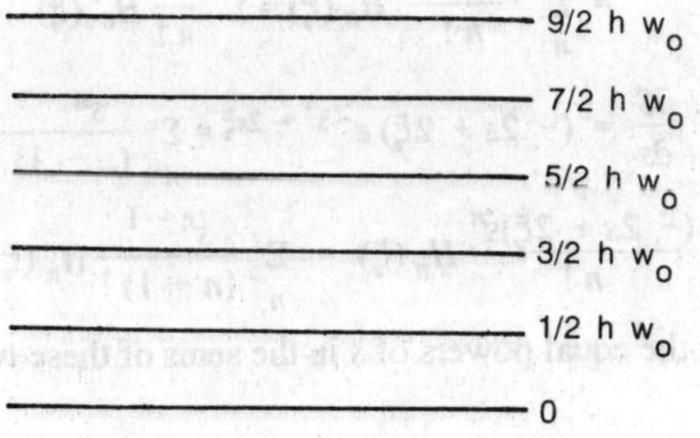

Fig. 4.2 *Energy level diagram for a linear harmonic oscillator.*

Hermite Polynomials

When $\lambda = (2n + 1)$, Eq. (4.38) will become

$$H_n'' - 2\,\xi H_n' + 2n\,H_n = 0 \qquad ...(4.45)$$

The polynomials $H_n(\xi)$ are called the Hermite polynomials. By using a convenient formulation we can express $H_n(\xi)$ in terms of a generating function $S(\xi, s)$

$$S(\xi, s) = e^{+\xi^2 - (s-\xi)^2} = e^{-s^2 + 2s\xi}$$

$$= 1 - s^2 2s\xi + \frac{(-s^2 + 2s\xi)^2}{2!} + \ldots$$

$$= 1 + 2s\xi + (4\xi^2 - 2)\frac{s^2}{2!} + \ldots$$

$$= H_0 + H_1 s + H_2 \frac{s^2}{2!} + \ldots$$

$$= \sum_{n=0}^{\infty} \frac{H_n(\xi) s^n}{n!} \quad \ldots(4.46)$$

Hence $$e^{-s^2 + 2s\xi} = \sum_{n=0}^{\infty} \frac{H_n(\xi) s^n}{n!} \quad \ldots(4.47)$$

To prove that $H_n(\xi)$ will satisfy the differential equation (4.45), both sides of Eq. (4.47) are differentiated *first* with respect to ξ and then with respect to s.

$$\frac{\partial S}{\partial \xi} = 2e^{-s^2 + 2s\xi} = \sum_n \frac{s^n}{n!} H_n'(\xi)$$

$$= \sum_n \frac{2s^{n+1}}{n!} H_n(\xi) = \sum \frac{s^n}{n!} H_c'(\xi)$$

$$\frac{\partial S}{\partial s} = (-2s + 2\xi)e^{-s^2 + 2s\xi} = \sum_n \frac{s^{n-1}}{(n-1)!} H_n(\xi)$$

$$\sum_n \frac{(-2s + 2\xi)s^n}{n!} H_n(\xi) = \sum_n \frac{s^{n-1}}{(n-1)!} H_n(\xi)$$

Equating the equal powers of s in the sums of these two equation, we get

$$H_n' = 2nH_{n-1} \quad \ldots(4.48)$$

$$H_{n+1} = 2\xi H_n - 2nH_{n-1} \quad \ldots(4.49)$$

Differentiating equations (4.48) and (4.49) with respect to ξ, we get

$$H_n'' = 2nH'_{n-1}$$

$$H'_{n+2} = 2\xi H_n' + 2nH'_{n-1} + 2H_n$$

Adding these, we get

$$H_n'' + H'_{n+1} = 2\xi H' + 2H_n \quad ...(4.50)$$

But $H'_{n+1} = 2(n+1)H_n$ [From equation (4.48)].

If the values of H'_{n+1} are substituted in equation (21), we get

$$H_n'' + 2nH_n + 2H_n = 2\xi H_n' + H_n$$

or

$$H_n'' - 2\xi H'_n + 2nH_n = 0.$$

Since H_n satisfies equation (4.45), so $H_n(\xi)$ are Hermite Polynomials.

When $S(\xi, s)$ has been differentiated n times with respect to s and s is then set equal to zero, Eq. (4.47) shows that the result would be simply $H_n(\xi)$. For any function of the form $f(s, \xi)$, it is clear that

$$\frac{\partial f}{\partial s} = -\frac{\partial f}{\partial \xi}$$

Now $$H_n(\xi) = \frac{\partial^n S}{\partial s^n}(\xi, s)$$

$$\frac{\partial^n}{\partial s^n}(e^{-s^2 + 2s\xi}) = \frac{\partial^n}{\partial s^n}[e^{\xi^2 - (s-\xi)^2}]$$

$$= e^{-\xi^2}\frac{\partial^n}{\partial s^n}e^{-(s-\xi)}$$

$$= (-1)^n e^{\xi^2}.\frac{\partial^n}{\partial \xi^v}e^{-(s-\xi)^2}$$

The above equation represents an expression for the nth Hermite Polynomial

$$H_n(\xi) = (-)^n e^{\xi^2}\frac{\partial^n}{\partial \xi^v}e^{-(\xi)^2} \text{ for } s = 0 \quad ...(4.51$$

The first few Hermite Polynomials are given below.

$H_1(\xi) = 1$, $H_1(\xi) = 2\xi$, $H_2(\xi) = 4\xi^2 - 2$, $H_2(\xi) = 8\xi^3 - 12h$;

$H_4(\xi) = 16\xi^4 - 48\xi - 12$.

The Series Solution

Let $$\psi = a_0 + a_1x + a_2x^2 + a_3x^3 + \ldots$$

$$= \Sigma a_n x^n = f(x)$$

Then the general solution is

$$\psi = e^{-\alpha x^2/2} f(x)$$

$$= e^{-\alpha x^2/2} \cdot \Sigma\, a_n x^n \qquad \text{...(4.52)}$$

For simplicity, Let

$$\sqrt{\alpha \cdot x = z}$$

Then the wave equation will be

$$\frac{d^2\psi}{dz^2} + \left(\frac{\lambda}{\alpha} - z^2\right)\psi = 0$$

and so $\psi = e^{-z^2/2} - f(x)$

$$= e^{-z^2/2} \cdot f\left(\frac{z}{\sqrt{\alpha}}\right)$$

$$= e^{-z^2/2}\, \phi(z)$$

where $\phi(z) = f(z/\sqrt{\alpha})$ ...(2)

If $\phi(z)$ is a power series in z, it can be written as

$$\phi(z) = A_0 + A_1 Z + A_2 Z_2 + \ldots$$

$$= \sum_0 A_n Z^n$$

Hence $\psi = e^{-2z/2} \sum_0 A_n z^n$ ...(3)

Differentiating eq. (4.53), we get

$$\frac{d\psi}{d^2} = e^{-z^2} \cdot \sum_0 n\, A_n\, Z^{n-1} + \left(-22 \cdot \frac{1}{2}\right) e^{-z^2/2}\, \sum_0 A_n Z|^n$$

$$= e^{-z^2/2}\left[\sum_0 n\, An\, Z^{n+1} - Z \sum_0 A\, n\, Z^n\right]$$

and

$$\frac{d^2\psi}{dz^2} = e^{-z^2/2} \sum_0 n\, An\, Z^{n-2} - \sum_0 (2n+1) A_n Z^n + \sum_0 n[n-1) A_n Z^{n-2}]$$

Hence the wave equation is

$$e^{-z^2}/2\left[\Sigma\, A_n\, Z^{z+2} - \Sigma(2n+1) A_n Z^n + \Sigma n(-1) A_n Z^{n-2}\right]$$

or

$$+ \left(\frac{\lambda}{\alpha} - z^2\right) z^{2/2} \cdot \Sigma\, A_n Z^n = 0$$

$$-\Sigma(2n+1) A_n Z^n + \Sigma n(n-1) A_n Z^{n-2} + \frac{\lambda}{\alpha} \Sigma A_n Z^n = 0.$$

As it is true for all values of x, so the coefficients of all powers of x, must varnish because z can not varnish. Thus equating the coefficients of Z_n to zero, we get

$$-A_n\left(2n+1-\frac{\lambda}{\alpha}\right)+(n+2)(n+1)A_{n+2}=0$$

or $$\frac{A_{n+2}}{A_n}=\frac{(2+1)-\frac{\lambda}{\alpha}}{(n+1)(n+2)} \qquad ...(4.55)$$

This formula is called the *recursion formula.*

The series consists of either odd or even powers of z. As this series represents the values of ψ so it is a convergent series and must terminate at some term beyond which the coefficients of higher powers must vanish.

Let it terminate at the $\frac{nth}{n}$ term then the coefficient of $(n+2)\frac{th}{n}$ term must vanish *i.e.,*, $A_{n+2}=0$.

or $$(2n+1)-\frac{\lambda}{\alpha}=0 \qquad ...(4.56)$$

or $$\frac{\lambda}{\alpha}=(2n+1)$$

Substituting for λ and α, we get

$$En=\left(n+\frac{1}{2}\right)hV \qquad ...(4.57)$$

where n may be zero or integer. From this equation we conclude the Schrodinger's equation for a linear harmonic oscillator can have physically acceptable solutions for the discrete values of $n = 0, 1, 2, \ldots$

From equation (5) it is apparent that for $n = 0_0$ nE_0 will be equal to $\frac{1}{2}hV$. Thus the lowest possible energy of a simple harmonic oscillator will be $\frac{1}{2}hV$ and this is called *zero point energy.*

At absolute zero when all other internal motions have ceased, the particle still possesses some vibrational energy which indicates that some atoms are still vibrating. This means that it is not possible to determine the precise position of the oscillator. This conclusion is in agreement with Heisenberg's uncertainty principle. Momentum in such cases is completely defined by the fact that $p = 0$.

The Eigenfunctions

In the previous article we have seen that

$$\psi = e^{-\alpha^2/2}\ \Sigma\ a_n x^n$$

$$= e^{-\alpha x^2/2}(a_0 + a_2 x^2 + a_4 x^4 + \ldots)$$

$$+ e^{-\alpha x^2/2}(a_1 x + a_3 a^3 + a_5 x^5 + \ldots)$$

Then for $n = 0$, we have

$$\psi_0 = a_0 \cdot e^{-\alpha x^2/2}$$

and for $n = 1$

$$\psi_1 = a_1 \cdot e^{-\alpha x^2/2}$$

Similarly for $n = 2$, $n = 3$ etc. we can find values of ψ.

Each of these eigenfunctions can be normalised and must be orthogonal, so that

$$\int_{-\alpha}^{\alpha} \psi_0{}^2\, dx = a_0{}^2 \int_{-\alpha}^{+\alpha} e^{-\alpha x^2}\, dx = 1$$

or

$$a_0{}^2 \sqrt{\frac{\pi}{\alpha}} = 1$$

or

$$a_0 = \left(\frac{\alpha}{\pi}\right)^{\frac{1}{4}}$$

$$\therefore \quad \psi_0 = \left(\frac{\alpha}{\pi}\right)^{\frac{1}{4}} e^{\alpha x^2/2}$$

Similarly

$$\psi_1 = \left(\frac{4\alpha^3}{\pi}\right)^{\frac{1}{4}} e^{-\alpha x^2/2}$$

$$\psi_2 = \left(\frac{\alpha}{4\pi}\right)^{\frac{1}{4}} (1 - 2\alpha x^2)\ e^{\alpha - x^2/2}$$

$$\psi_3 = \left(\frac{9\alpha^3}{\pi}\right)^{\frac{1}{4}} \left(x - \frac{z\alpha}{3} x^3\right) e^{-\alpha x^2/2}$$

and so on

The Recursion Formula for the Hermite Polynomials

On introducing Eq. $\psi_n(\xi) = \exp(-\xi^2/2)\, H_n(\xi)$ into Equation

$$\frac{\partial^2\psi}{\partial\xi^2} + \left(\frac{\alpha}{\beta} - \xi^2\right)\psi = 0$$

we get the following differential equation for the Hermite functions

$$\frac{\partial^2 H_n}{\partial\xi^2} - 2\xi\frac{\partial H_n}{\partial\xi} + \left(\frac{\alpha}{\beta} - \xi^2\right) H_n(\xi) + (\xi^2 - 1) H_n = 0 \quad ...(4.58)$$

Inserting Eq. $\frac{\alpha}{\beta} = 2n + 1$ in Eq. (1) we get

$$\frac{\partial^2 H_n}{\partial\xi^2} - 2\xi\frac{\partial H_n}{\partial\xi} + 2n\, H_n = 0 \quad ...(4.59)$$

Differentiating twice the Eq.

$$H_n(\xi) = (-1)^n \exp(\xi^2)\frac{\partial^n}{\partial\xi^n}[\exp(\xi)^2]$$

we have

$$\frac{\partial^2 H_n}{\partial\xi^2} = 2_n + 2\xi\frac{\partial H_n}{\partial\xi} - \frac{\partial H_{n+1}}{\partial\xi} \quad ...(4.60)$$

Eliminating $\frac{\partial^2 Hn}{\partial\xi^2}$ from Eqs. (4.59) and (4.60) we get

$$2\xi\frac{\partial H_n}{\partial\xi} 2nHn = 2Hn + 2\,\xi\frac{\partial Hn}{\partial\xi} - \frac{\partial H_{n+1}}{\partial\xi} \quad ...(4.61)$$

which becomes

$$\frac{\partial H_{n+1}}{\partial\xi} = 2(n+1) H_n \quad ...(4.62)$$

Since

$$H_{n+1}(\xi) = (-)^{n+1}\exp(\xi^2/2)\frac{\partial^{n+1}}{\partial\,\xi^{n+1}}\exp(-\xi^2) \quad ...(4.63)$$

$$\frac{\partial H_{n+1}}{\partial\xi} = 2\xi H_{n+1} - H_{n+2} \quad ...(4.64)$$

From Eqs. (4.62) and (4.64) we get

$$H_{n+2} = 2\xi H_{n+1} - 2(n+1) Hn \quad (4.65)$$

This relation will give the higher polynomials. On replacing $(n + 1)$ by n', another integer, we obtain

$$\xi Hn' = n'Hn'_{-1} + \frac{1}{2}Hn'_{+1} \quad ...(4.66)$$

which is the required recursion formula for the Hermite polynomials.

The Selection Rules for the Harmonic Oscillator

Since we can have an infinite number of energy level for a harmonic oscillator so it would seem that the spectrum might be very complex. However observe simplicity of the molecular vibration spectrum may be due to the operation of 'selection rules' that govern the intensities of the lines or bands which are associated with many transitions which might take place. It is possible to derive the selection rules by examining the transition dipole moment. A transition between two states characterized by the vibrational quantum numbers n and n' is possible only if the dipole moment of transition is not equal to zero, *i.e.*

$$e\int_{-\infty}^{\infty} \psi n^* \, X \, \psi n \, dx \neq 0 \quad ...(4.67)$$

Substituting ψn and $\psi n'$ and using $\xi = \sqrt{\beta}\, x$ we get

$$e\int_{-\infty}^{\infty} \psi n^* \cdot \hat{X}\, \psi\, n' \, dx = \frac{e}{\beta}\left[\frac{1}{2^n n!\sqrt{\pi}}\right]^{1/2} \times$$

$$\left[\frac{1}{2^n n'!\sqrt{\pi}}\right]^{1/2} \int_{-\infty}^{+\infty} \exp\,(-\xi^2)\, Hn \cdot \xi H_n'\, d\xi \quad ...(4.68)$$

$$\int_{-\infty}^{\infty} \exp.\,(-\xi^2)\, Hn\, \xi Hn'\, d\xi$$

$$= \int_{-\infty}^{\infty} \exp.\,(-\xi^2)\, H_n\,(n'H^{n-1} + \frac{1}{2}H^{n+1})\, d\xi$$

$$= (N^n N^{n-1})^{-1} \int_{-\infty}^{\infty} \xi^* n \xi n'_{-1}\, d\xi + \frac{3}{4}(N^n N_{n+1})^{-1}$$

$$\int_{-\infty}^{\infty} \psi^{n*} \psi_{n+1} \, d\xi \qquad ...(4.69)$$

where $(Nn')\, Nn_{-1}$ and Nn'_{+1} are the normalisation factors for the states ψ_n, and $\psi n'_{+1}$ respectively. Because wave functions for different states of a harmonic oscillator are mutually orthogonal so we can say that the transition dipole moment integral is zero unless $n = n' \pm 1$. The selection rule must be

$$\Delta n = \pm 1 \qquad ...(4.70)$$

For a harmonic oscillator, therefore, only the fundamental frequency, V_0, is emitted or absorbed. If the oscillator is somewhat enharmonic, this selection rule is not obeyed and the oscillator can then absorb or emit radiations of frequencies v_0, $2v_0$, $3v_0$.....and so on. The intensities of overtone bands give an indication of the amount of enharmonicity the vibration.

5

Various Solutions of Schrodinger's Equation for Hydrogen Like Atoms

Introduction

THE HYDROGEN ATOM

No matter how far quantum mechanics might go towards this end, a practical. It is generally possible to set up the differential equation for a particular problem but it is found that the resulting differential equation in most cases can rarely be solved without resorting to approximation methods. In fact, only a few quantum mechanical problems can be solved without some form of approximation, and the hydrogen-like atom is one of these. This fact alone is sufficient to emphasize the importance of the hydrogen atom problem.

The problem, of course, is to solve the Schroedinger equation as set up, for the hydrogen atom. The Schroedinger's wave equation in the form

$$\nabla^2 \psi_{(x\,y\,z)} + \frac{8\pi^2 m}{h^2}(E - V)\,\psi_{(x\,y\,z)} = 0$$

has proved satisfactory for the motion of a single particle of mass m. However, in the hydrogen atom there are two particles, the electron and the nucleus. For such a system the wave equation can be conveniently put in the form

$$\frac{1}{m}\nabla^2 \psi_{(x\,y\,z)} + \frac{8\pi^2}{h^2}(E - V)\,\psi_{(x\,y\,z)} = 0$$

Now if, we consider the motion of the two particles in the hydrogen atom, the wave equation becomes

$$\frac{1}{m_1}\nabla_1{}^2\psi_T + \frac{1}{m^2}\nabla_2{}^2\psi_T + \frac{8\pi^2}{h^2}\left(E + \frac{e^2}{r}\right)\psi_T = 0$$

Transformation of Coordinates

We find that the total energy, *E*, in the wave equation is made up of two parts : (1) the translational motion of the atom as a whole, and (2) the energy of the electron with respect to the proton. At present our interest is mainly in the latter portion of the energy. This leads us to the problem of separation of variables. To obtain the desired equation, it is essential to separate out and discard the translational portion of the total wave equation. This can be achieved if we introduce a new set of variables *x, y* and *z*, which are Cartesian coordinates of the *centre of mass* of the hydrogen atom, and the variables *r*, θ and φ, which are polar coordinates of the electron with respect to the nucleus.

A coordinate of the centre of mass of a system is, generally given by

$$q = \frac{\sum_i m_i q_i}{\sum_i m_i}$$

where m_i is the mass of the ith particle and q_i is the q coordinate of the ith particle. For the hydrogen atom, the cartesian coordinates of the centre of mass will be given by

$$x = \frac{m_1x_1 + m_2x_2}{m_1 + m_2} \qquad ...(5.1)$$

$$y = \frac{m_1y_1 + m_2y_2}{m_1 + m_2} \qquad ...(5.2)$$

$$z = \frac{m_1z_1 + m_2z_2}{m_1 + m_2} \qquad ...(5.3)$$

and the transformation to spherical coordinates can be seen from Fig. 7.1 to be

$$r \sin\theta \cos\varphi = x_2 - x_1 \qquad ...(5.4)$$

$$r \sin\theta \sin\varphi = y_2 - y_1 \qquad ...(5.5)$$

$$r \cos\theta = z_2 - z_1 \qquad ...(5.6)$$

Making use of these transformation equations we can obtain the wave equation in terms of the Cartesian coordinates of the centre of mass of the system and the polar coordinates r, θ and φ. The x, y, z coordinates of the centre of mass of the atom relate to the translational motion of the atom as a whole, and the r, θ, φ coordinates relate to the coordinates of the electron (x_1, y_1, z_1) to the coordinates of the nucleus (x_2, y_2, z_2).

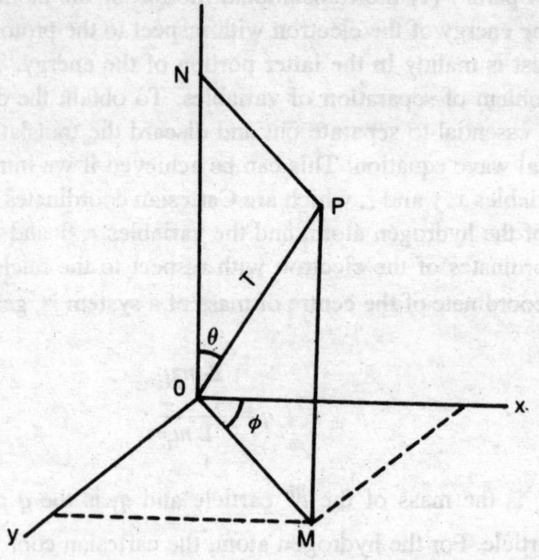

Fig. 5.1. ***Transformation Diagram for Spherical Coordinates.***

For example solving Eq. (5.3) for z_2, it can be seen that

$$z_2 = \left(\frac{m_1 + m_2}{m_2}\right) z - \frac{m_1}{m_2} z_1$$

If the value of z_2 is now put into Eq. (5.6), it can be found that

$$r \cos\theta = \left(\frac{m_1 + m_2}{m_2}\right) z - \frac{m_1}{m_2} z_1 - z_1$$

or

$$z_1 = z - \left(\frac{m_2}{m_1 + m_2}\right) r \cos\theta$$

multiplying by the term m_1/m_2, we obtain

$$z_1 = z - \frac{\mu}{m_1} r \cos\theta \qquad ...(5.7)$$

where μ is the reduced mass of the system and is given by

$$\mu = \frac{m_1 m_2}{m_1 + m_2} \qquad ...(5.8)$$

Following this procedure, a transformation equation can be found for each of the coordinates, and by making proper substitutions the wave equation can be obtained in terms of the new variables x, y, z, r, θ and φ. Although the procedure is tedious, yet it is straightforward. In terms of the new variables, it can be found tobe

$$\frac{1}{m_1 + m_2}\left(\frac{\partial^2 \psi_T}{\partial x^2} + \frac{\partial^2 \psi_T}{\partial y^2} + \frac{\partial^2 \psi_T}{\partial z^2}\right) + \frac{m_1 + m_2}{m_1 m_2}\left\{\frac{1}{r^2}\frac{\partial}{\partial r}\left(r^2 \frac{\partial \psi_T}{\partial T}\right)\right.$$

$$\left. + \frac{1}{r^2 \sin^2\theta}\frac{\partial^2 \psi_T}{\partial \psi^2} + \frac{1}{r^2 \sin\theta}\frac{\partial}{\partial \theta}\left(\sin\theta \frac{\partial \psi_T}{\partial \theta}\right)\right\}$$

$$+ \frac{8\pi^2}{h^2}[E - V_{(r)\,T}]\,\psi = 0 \qquad ...(5.9)$$

The wave function, ψ_T, is a function of the variables x, y, z, r, θ, and φ, and the energy, E, contains the translational energy of the atom as also the energy of the electron with respect to the proton.

The aim of this transformation to new coordinates is to make a separation of variables possible. In principles, the separation can be done in the same way as it was in the particle in a box problem. However, in this particular case, the algebra is a bit more complex. In the usual manner, the total wave function, $\psi_{(x\,y\,z\,r\,\theta\,\varphi)}$, can be assumed to be assumed to be expressible as the product of two wave functions such that

$$\psi_{(x\,y\,z\,r\,\theta\,\varphi)} = F_{(x\,y\,z)}\,\psi_{(r\,\theta\,\varphi)}$$

substituting it into Eq. (5.9), the following two equations can be obtained.

$$\frac{\partial^2 F_{(x\,y\,z)}}{\partial x^2} + \frac{\partial^2 F_{(x\,y\,z)}}{\partial y^2} + \frac{\partial^2 F_{(x\,y\,z)}}{\partial z^2}$$

$$+ \frac{8\pi^2 (m_1 + m_2)}{h^2} E_{trans} F_{(x\,y\,z)} = 0 \qquad ...(5.10)$$

$$\frac{1}{r^2}\frac{\partial}{\partial r}\left(r^2\frac{\partial \psi_{(r\theta\varphi)}}{\partial r}\right)+\frac{1}{r^2\sin^2\theta}\frac{\partial^2\psi_{(r\theta\varphi)}}{\partial\psi^2}+\frac{1}{r^2\sin\theta}\frac{\partial}{\partial\theta}$$

$$\left(\sin\theta\frac{\partial\,\psi_{(r\theta\varphi)}}{\partial\theta}\right)+\frac{8\pi^2\mu}{h^2}[E-V_{(r)}]\,\psi_{(r\theta\varphi)}=0 \quad ...(5.11)$$

The first of these equations contains only the variables x, y and z, and has no potential energy term. And is identical to the wave equation for a free particle, so it represents the translational energy of the atom as a whole. The second equation, relates the electron to the proton, and is the equation of particular interest to us.

Separation of Variables

As we are interested in the second part of the total wave equation so we will discard the translational part. The desired wave equation for the electron with respect to the nucleus. Then is a second-order partial differential equation, and the standard methods must be used to get its solution. For this the variables be separated in such a way that three independent equations are obtained, each containing only one of the three variables.

To separate the variables, it is essential to assume that the wave function $\psi_{(r\theta\varphi)}$ can be represented by the product of three wave functions, each containing one and only one of the three variables r, θ, and φ. If we let

$$\psi_{(r\theta\varphi)}=R_{(r)}\,\Theta_{(\theta)}\,\Phi_{(\varphi)}$$

and make this substitution into Eq. (11), we obtain

$$\frac{1}{r^2}\frac{\partial}{\partial r}\left(r^2\frac{\partial}{\partial r}R_{(r)}\,\Theta_{(\theta)}\,\Phi_{(\varphi)}\right)+\frac{1}{r^2\sin^2\theta}\frac{\partial^2}{\partial_\varphi^{\,2}}R_{(r)}\,\Theta_{(\theta)}\,\Phi_{(\varphi)}$$

$$+\frac{1}{r^2\sin\theta}\frac{\partial}{\partial\theta}\left(\sin\theta\frac{\partial}{\partial\theta}R_{(r)}\,\Theta_{(\theta)}\,\Phi_{(\varphi)}\right)$$

$$+\frac{8\pi^2\mu}{h^2}[E-V_{(r)}]\,R_{(r)}\,\Theta_{(\theta)}\,\Phi_{(\varphi)}=0 \quad ...(5.12)$$

dividing by $R_{(r)}\,\Theta_{(\theta)}\,\Phi_{(\varphi)}$ Eq. (5.12), we get

$$\frac{1}{r^2R_{(r)}}\frac{\partial}{dr}\left(r^2\frac{\partial R_{(r)}}{\partial r}\right)+\frac{1}{\Phi_{(\varphi)}\,r^2\sin^2\theta}\frac{\partial^2\Phi_{(\varphi)}}{\partial_\varphi^{\,2}}$$

$$+\frac{1}{\Theta_{(\theta)}\, r^2 \sin\theta}\frac{\partial}{\partial\theta}\left(\sin\theta\frac{\partial\,\Theta_{(\theta)}}{\partial\theta}\right)+\frac{8\pi^2\mu}{h^2}[E-V_{(r)}]=0 \quad ...(5.13)$$

Multiply Eq. (5.13) by $r^2 \sin^2\theta$, we get

$$\frac{\sin^2\partial}{R_{(r)}}\frac{\partial}{\partial r}\left(r^2\frac{\partial R_{(r)}}{\partial r}\right)+\frac{1}{\Phi_{(\varphi)}}\frac{\partial^2\Phi_{(\varphi)}}{\partial_\varphi{}^2}+\frac{\sin\theta}{\Theta_{(\theta)}}\frac{\partial}{\partial\theta}\left(\sin\theta\frac{\partial\,\Theta_{(\theta)}}{\partial\theta}\right)$$

$$+\frac{8\pi^2\mu\, r^2\sin^2\theta}{h^2}[E-V_{(r)}]=0 \quad ...(5.14)$$

or

$$\frac{\sin^2\theta}{R_{(r)}}\frac{\partial}{\partial r}\left(r^2\frac{\partial R_{(r)}}{\partial r}\right)+\frac{\sin\theta}{\Theta_{(\theta)}}\frac{\partial}{\partial\theta}\left(\sin\theta\frac{\partial\,\Theta_{(\theta)}}{\partial\theta}\right)$$

$$+\frac{8\pi^2\mu\, r^2\sin^2\theta}{h^2}[E-V_{(r)}]=-\frac{1}{\Phi_{(\varphi)}}\frac{\partial^2\Phi_{(\varphi)}}{\partial_\varphi{}^2} \quad ...(5.15)$$

This is a situation that is analogous to that which arose for the particle in a three-dimensional box. The left side of Eq. (5.15) contains only the variables r and θ and the right side of the equation contains only variable Φ. No matter what values r, θ, or Φ might independently take, the sum of the terms on the left must always equal the term on the right. This can be true only if each side of the equation is equal to the same constant. If this constant be m^2 then it can be seen that the variable φ can immediately be separated from Eq. (5.15), giving

$$\frac{1}{\Phi_{(\varphi)}}\frac{d^2\Phi_{(\varphi)}}{d_\varphi{}^2}=-m^2 \quad ...(5.16)$$

Hence, we have succeeded in separating the first of the three variables from the original wave equation. The problem now is to carry out the separation of the remaining two variables r and θ. This can be achieved by equating the portion of the equation containing $R_{(r)}$ and $\Theta_{(\theta)}$ to the constant m^2, so that

$$\frac{\sin^2\theta}{R_{(r)}}\frac{\partial}{\partial r}\left(r^2\frac{\partial R_{(r)}}{\partial r}\right)+\frac{\sin\theta}{\Theta_{(\theta)}}\frac{\partial}{\partial\theta}\left(\sin\theta\frac{\partial\,\Theta_{(\theta)}}{\partial\theta}\right)$$

$$+\frac{8\pi^2\mu^2\sin^2\theta}{h^2}[E-V_{(r)}]=m^2 \qquad ...(5.17)$$

Dividing Eq. (5.17) by $\sin^2\theta$, we get

$$\frac{1}{R_{(r)}}\frac{\partial}{\partial r}\left(r^2\frac{\partial R(r)}{\partial r}\right)+\frac{1}{\Theta_{(\theta)}\sin\theta}\frac{\partial}{\partial\theta}\left(\sin\theta\frac{\partial\Theta_{(\theta)}}{\partial\theta}\right)-\frac{m^2}{\sin^2\theta}$$

$$+\frac{8\pi^2\mu r^2}{h^2}[E-V_{(r)}]=0 \qquad ...(5.18)$$

This can be written as

$$\frac{1}{R_{(r)}}\frac{\partial}{\partial r}\left(r^2\frac{\partial R_{(r)}}{\partial r}\right)+\frac{8\pi^2\mu r^2}{h^2}[E-V_{(r)}]$$

$$=\frac{m^2}{\sin\theta^2}-\frac{1}{\theta_{(r)}\sin}\frac{\partial}{\partial\theta}\left(\sin\theta\frac{\partial\Theta_{(r)}}{\partial\theta}\right) \qquad ...(5.19)$$

Since each side of the above equations contains only one so they both must be equal to the same constant. If the right side of the equation is set equal to the constant β, this gives on multiplication by θ.

$$\frac{m^2\theta_{(\theta)}}{\sin^2\theta}-\frac{1}{\sin\theta}\frac{d}{d\theta}\left(\sin\theta\frac{d\theta_{(\theta)}}{d\theta}\right)-\beta\theta_{(\theta)}=0 \qquad ...(5.20)$$

This is the desired form of the Θ equation. The remaining part of the original equation is the R equation,

$$\frac{1}{r^2}\frac{d}{dr}\left(r^2\frac{dR_{(r)}}{dr}\right)-\frac{\beta}{r^2}R_{(r)}+\frac{8\pi^2}{h^2}[E-V_{(r)}]R_{(r)}=0 \qquad ...(5.21)$$

Thus, the three variables have been successfully separated, and the three independent total differential equations obtained are

$$\frac{d^2\Phi(\varphi)}{d_\varphi}+m^2\Phi(\varphi)=0 \qquad ...(5.22)$$

$$\frac{1}{\sin\theta}\frac{d}{d\theta}\left(\sin\theta\frac{d\Theta_{(\theta)}}{d\theta}\right)-\frac{m^2\Theta_{(\theta)}}{\sin^2\theta}+\beta\Theta_{(\theta)}=0 \qquad ...(5.23)$$

$$\frac{1}{r^2}\frac{d}{dr}\left(r^2\frac{dR_{(r)}}{dr}\right)-\frac{\beta}{r^2}R_{(r)}+\frac{8\pi^2\mu}{h^2}[E-V_{(r)}]R_{(r)}=0 \qquad ...(5.24)$$

The Φ Equation

The first of these equations is known as the **Φ equation,** and it is of the same form as the wave equation for the particle in a box. In terms of since and cosine, the solution is

$$\Phi_m(\varphi) = A \sin m_\varphi + B \cos m_\varphi \quad \text{...(5.25)}$$

For a wave function to be acceptable, it must be a well-behaved function. The requirement of such a function is that it should be *single valued.* For this the function $\Phi_m(\varphi)$ must have the same value for $_\varphi = 0$ as it does for $_\varphi = 2\pi$. For the case of $_\varphi = 0$, it can be seen that

$$\Phi_m(0) = A \sin(0) + B \cos(0) = B \quad \text{...(5.26)}$$

and when $\varphi = 2\pi$, we have

$$\Phi_m(2\pi) = A \sin m2\pi + B \cos m2\pi \quad \text{...(5.27)}$$

Since the value of $\Phi_m(\varphi)$ must be the same under both of these conditions, therefore,

$$B = A \sin m2\pi + B \cos m2\pi \quad \text{...(5.28)}$$

This identity can be true only if *m* is zero or has a positive or negative integral value. These characteristics are similar to those expected of a quantum number, and the particular restrictions on *m* show that it is the analog of the magnetic quantum number of the Bohr-Sommerfeld model.

Generally, in the treatment of the hydrogen atom, the exponential solution to the Φ equation,

$$\Phi_m(\varphi) = Ce^{\pm im_\varphi}$$

is used. For evaluation of the constant, *C*, it would be convenient to choose *C* in such a way that the wave function, $\Phi_{(}\varphi)$, becomes a normalized wave function. This means that

$$\int_0^{2\pi} \Phi\Phi^* \, d_\varphi = 1 \quad \text{...(5.29)}$$

which gives

$$\int_0^{2\pi} C^2 e^{im_\varphi} \; e^{\pm im_\varphi} d_\varphi = 1 \quad \text{...(5.30)}$$

or

$$C^2 \int_0^{2\pi} d_\varphi = 2\pi C^2 = 1 \quad \text{...(5.31)}$$

The value of the constant, C which gives a normalized wave function is given by

$$C = 1/\sqrt{2\pi},$$

and the final normalized wave function is given by

$$\Phi_m(\varphi) = \frac{1}{\sqrt{2\pi}} e^{\pm im_\varphi}, \quad m = 0, \pm 1, \pm 2, \ldots \qquad \text{...(5.32)}$$

The Hydrogen-like Atom

Once again we consider the hydrogen like atom and consider it as a system of two interacting point particles, the nature of interaction between them is of Coulobmic attraction of their electric charge. Let the nucleus possess a charge $+ Ze$ where Z is the atomic number of the element and let $- e$ be the charge on the electron. Thus the potential energy of the system is $-\frac{Ze^2}{r}$, where r is the distance between the nucleus and the electron. The wave equation for the electron is given by

$$\nabla^2\psi + \frac{8\pi^2 m}{h^2}\left(E + \frac{Ze^2}{r}\right)\psi = 0 \qquad \text{...(5.33)}$$

Transforming into polar co-ordinates we get

$$\frac{1}{r^2}\frac{d}{dr}\left(r^2\frac{d\psi}{dr}\right) + \frac{1}{r^2 \sin\theta}\frac{d}{d\theta}\left(\sin\theta\frac{d\psi}{d\theta}\right) +$$

$$\frac{1}{r^2 \sin^2\theta}\frac{d^2\psi}{d\phi^2} + \frac{8\pi^2 m}{h^2}\left(E + \frac{Ze^2}{r}\right)\psi = 0 \qquad \text{...(5.34)}$$

Assuming that the eigenfunctions ψ is a function of r, θ and ϕ can be expressed as the product of three functions :

$$\psi(r, \theta, \phi) = R(r), \Theta(\theta), \Phi(\phi)$$

where $R(r)$ or R is a function of r only, $\Theta(\theta)$ or simply Θ, is a function of θ only and $\Phi(\phi)$ or Φ that of ϕ only.

Substituting this value of ψ in (5.34) we get

$$\frac{\Phi\Theta}{r^2}\frac{d}{dr}\left(r^2\frac{dR}{dr}\right) + \frac{R\Phi}{r^2 \sin\theta}\frac{d}{d\theta}\left(\sin\theta\frac{d\Theta}{d\theta}\right) + \frac{R\Theta}{r^2 \sin^2\theta}\frac{d^2\Phi}{d\phi^2}$$

$$= \frac{8\pi^2 m}{h^2}\left(E + \frac{Ze^2}{r}\right)R\Theta\Phi = 0. \qquad \text{...(5. 35)}$$

Dividing by R $\Theta\Phi$ we get

$$\frac{1}{Rr^2}\frac{d}{dr}\left(r^2\frac{dR}{dr}\right)+\frac{1}{\Theta^2\sin\theta}\frac{d}{d\theta}\left(\sin\theta\frac{d\Theta}{d\theta}\right)+$$

$$\frac{1}{\Phi r^2}\frac{d^2\Phi}{\sin^2\theta\, d\phi^2}+\frac{8\pi^2 m}{h^2}\left(E+\frac{Ze^2}{r}\right)=0. \quad ...(5.36)$$

Multiplying by $r^2\sin^2\theta$ and rearranging

$$\frac{\sin^2\theta}{R}\frac{d}{dr}\left(r^2\frac{dR}{dr}\right)+\frac{\sin\theta}{\Theta}\frac{d}{d\theta}\left(\sin\theta\frac{d\Theta}{d\theta}\right)+\frac{1}{\Phi}\cdot\frac{d^2\Phi}{d\phi^2}$$

$$+\frac{8\pi^2 m}{h^2}\left(E+\frac{Ze^2}{r}\right)r^2\sin^2\theta=0. \quad ...(5.37)$$

Rearranging once again

$$\frac{\sin^2\theta}{R}\frac{d}{dr}\left(r^2\frac{dR}{dr}\right)+\frac{\sin\theta}{\Theta}\cdot\frac{d}{d\theta}\left(\sin\theta\frac{d\Theta}{d\theta}\right)+$$

$$\frac{8\pi^2 mr^2\sin^2\theta}{h^2}\left(E+\frac{Ze^2}{r}\right)=-\frac{1}{\Phi}\frac{d^2\Phi}{d\phi^2}=m^2 \quad ...(5.38)$$

Since R, Θ and Φ are all independent and a function of R and Θ can never be equal to one of Φ unless each is equal to a constant m^2 (say) as already seen

Thus $$\frac{d^2\Phi}{d\phi^2}+m^2\Phi=0. \quad ...(5.39)$$

Equation (5.38) can be written as

$$\frac{1}{R}\frac{d}{dr}\left(r^2\frac{dR}{dr}\right)+\frac{8\pi^2 mr^2}{h^2}\left(E+\frac{Ze^2}{r}\right)$$

$$=\frac{m^2}{\sin^2\theta}-\frac{1}{\Theta\sin\theta}\frac{d}{d\theta}\left(\sin\theta\frac{d\Theta}{d\theta}\right)$$

$$=\beta\ (\text{constant}) \quad ...(5.40)$$

since in this case the left hand side is a function of r only while the right hand side is a function of θ only, and they can never be related unless each is equal to a constant β. So we have

$$\frac{1}{\sin\theta}\frac{d}{d\theta}\left(\sin 0\frac{d\Theta}{d\theta}\right)+\left(\beta-\frac{m^2}{\sin^2\theta}\right)\Theta=0 \quad ...(5.41)$$

and
$$\frac{1}{R}\frac{d}{dr}\left(r^2\frac{dR}{dr}\right)+\frac{8\pi^2 mr^2}{h^2}\left(E+\frac{Ze^2}{r}\right)=\beta. \quad ...(5.42)$$

The initial wave equation is thus split up into three separate equations (5.39), (5.41) and (5.42). These equations can be solved to find the permitted values of energy. We shall first find out that equation (5.39) possesses acceptable solutions only for certain values of the parameter m. Then we shall introduce these values of m in equation (5.41) and find that equation (5.41) possesses solutions permissible for certain values of β. Finally we shall introduce these values of β in (5.42) and note that the equation possesses certain acceptable solutions only for certain values of E, which constitute the allowed values of energy.

Solution of Φ equation

The Φ equation is

$$\frac{d^2\Phi}{d\phi^2}+m^2\Phi=0 \quad ...(5.43)$$

or
$$\Phi = C.e.^{\pm im\varphi} \quad ...(5.44)$$

Taking $\Phi = C.e.^{+im\phi}$. Since Φ must be single-valued it must have the same value when $\phi = 0$ and $\phi = 2\pi$ *i.e.* after a complete rotation of the system.

When $\phi = 0,\quad \Phi = C$

and $\phi = 2\pi \quad \Phi = Ce^{2\pi im}$

Therefore $e^{2\pi im} = 1$

or
$$\cos 2\pi m + i\sin 2\pi m = 1.$$

This can be true only when 'm' is either = 0 or an integer. Since the rotation can occur either in the positive or negative direction m may be positive or negative. Hence the various permissible values of m are $m = 0, \pm1, \pm2, \pm3$ etc.

$$\frac{1}{\sin\theta}\frac{d}{d\theta}\left(\sin\theta\frac{d\theta}{d\theta}\right)\beta\left(\beta-\frac{m^2}{\sin^2\theta}\right)\theta=0. \quad ...(5.46)$$

Let $\cos\theta = \mu$, where μ lies between + 1 and - 1, and

$$d\mu = -\sin\theta\, d\theta$$

or
$$\frac{d}{d\theta} = -\sin\theta\frac{d}{d\mu}.$$

Then equation (5.46) becomes

$$\frac{d}{d\mu}\left[(1-\mu^2)\frac{d\theta}{d\mu}\right]+\left[\beta-\frac{m^2}{1-\mu^2}\right]\theta = 0 \qquad ...(5.47)$$

Let now $\theta = (1-\mu^2)^{m/2}\, F(\mu)$

or
$$\frac{d\theta}{d\mu} = (1-\mu^2)^{m/2}\frac{dF}{d\mu} + \left[F\frac{d(1-\mu^2)^{m/2}}{a(1-\mu^2)}\cdot\frac{d(1-\mu^2)}{d\mu}\right]$$

$$= (1-\mu^2)^{m/2}\frac{dF}{d\mu} - 2\mu F\cdot\frac{m}{2}(1-\mu^2)^{m/2-1}$$

$$= (1-\mu^2)^{m/2}\frac{dF}{d\mu} - mF\mu(1-\mu^2)^{m/2}-1 \qquad ...(5.48)$$

Substituting (48) in (46) we get

$$\frac{d}{d\mu}\left[(1-\mu^2)^{m/2+1}\frac{dF}{d\mu} - mF\mu\,(1-\mu^2)^{m/2}\right] + \beta F(1-\mu^2)^{m/2} - m^2(1-\mu^2)^{m/2-1}F = 0 \qquad ...(5.49)$$

or
$$(1-\mu^2)^{m/2+1}\frac{d^2F}{d\mu^2} - 2\mu^{(m/2+1)}(1-\mu^2)^{m/2}\frac{dF}{d\mu} - m\mu\,(1-\mu^2)^{m/2}\frac{dF}{d\mu} - mF(1-\mu^2)^{m/2} + 2m\cdot\frac{m}{2}.$$

$$F\mu^2(1-\mu^2)^{m/2-1} + \beta F(1-\mu^2)^{m/2} - m^2(1-\mu^2)^{m/2-1}F = 0 \qquad ...(5.50)$$

Dividing Eq. (5.50) by $(1-\mu^2)^{m/2}$ we get

$$(1-\mu^2)\frac{a^2F}{d\mu^2} - 2\mu\left(\frac{m}{2}+1\right)\frac{dF}{d\mu} - mF - m\mu\frac{dF}{d\mu} + \frac{m^2\mu^2}{1-\mu^2}F + \frac{m^2F}{1-\mu^2} + BF = 0 \qquad ...(5.51)$$

The θ Equation

The θ equation can be written as

$$\frac{\sin\theta}{\theta}\frac{d}{d\theta}\left(\sin\theta\frac{d\theta}{d\theta}\right) + \beta\sin^2\theta - m^2 = 0 \qquad ...(5.52)$$

The solutions to the θ equation as also of the radial equation, are not so simple as the solution to the Φ equation. However, the θ equation can easily be put into a form that was known to the mathematicians many years before the advent of quantum mechanics. In this particular form the equation is called **Legendre's equation** and has the normalized solution

$$\theta\, l_m(\theta) = \sqrt{\frac{(2l-1)(t-|m|)!}{2(l+|m|}}\; P_l(m)(\cos\theta) \qquad \text{...(5.53)}$$

where $P_l(m)$ is the associated Legendre function of degree l and order $|m|$. This solution is quite complicated, but for particular values of the parameters l and m the solution reduces to much simpler forms.

Some important features that can be observed are that $\beta = l(l+1)$ where the permitted values of l are 0, 1, 2, 3,... This is the source of the new parameter and its properties seem to be similar in many ways to those of the azimuthal quantum number of the Bohr-Sommerfeld atom. However, a new restriction is imposed on the quantum number m. In the normalizing factor of the solution to this equation the term $(l-|m|)!$ occurs. If $|m|$ is permitted to be greater than the factorial of a negative number results. Thus, the restrictions on the quantum number m now become $m = 0, \pm 1, \pm 2, \pm 3, ...l$. These are the same restrictions that were found to be necessary for the magnetic quantum number in the Bohr-Sommerfeld theory.

The θ equation [Eq. (5.45)] is

$$\frac{\sin\theta}{\theta}\frac{d}{d\theta}\left(\sin\frac{d\theta}{d\theta}\right) + \beta\sin^2\theta - m^2 = 0.$$

Dividing by

$$\frac{\sin^2\theta}{\theta},$$

we get

or $$(1-\mu^2)\frac{d^2F}{d\mu^2} - (m\mu + 2\mu + m\mu\frac{dF}{d\mu} +$$

$$\left[\beta - m - \frac{m^2\mu^2}{1-\mu^2} + \frac{m^2}{1-\mu^2}\right]F = 0 \qquad \text{...(5.54)}$$

or $$(1-\mu^2)\frac{d^2F}{d\mu^2} - 2\mu(m+1)\frac{dF}{d\mu} + (\beta - m - m^2)F = 0$$

...(5.55)

This homogeneous equation can now be solved by series method already explained.

The θ solution for ψ has been obtained as

$$= (-\mu^2)^{\frac{m}{2}} F(\mu) \qquad ...(5.56)$$

Substituting $\mu = \cos\theta$

$$\psi = \sin\theta^{\frac{m}{-}} F(\cos\theta) \qquad ...(5.57)$$

where c is a constant

Jeans proposed to represent the polynomial F (μ) as

$$F(\mu) = \frac{d^{l+m}}{d^{l}\mu^{+m}}(\mu^2 - 1)^l \qquad ...(5.58)$$

By carrying out successive differentiation of the above it can be shown that polynomial will terminate at

$$m = 1$$

If we represent

$$F(\mu = b_0 + b_1 \cos\theta + b_2 \cos^2\theta + \ldots \qquad ...(5.59)$$

then $\psi = c \cdot \sin^{|m|\theta}(b_0 + b_1 \cos\theta + b_2 \cos^2\theta + \ldots$

in which $b_0 \neq 0$. ...(5.60)

Combining this with θ solution, we get the total angular solution for ψ.

$$\psi = c \sin^{|m|\theta}(b_0 + b_1 \cos\theta + \ldots..) e^{im\phi} \qquad ...(5.61)$$

In it ψ has not been normalised.

We can represent ψ also as

$$\psi = c \sin^{|m|\theta} \frac{d^{l+m}}{d\mu^{l+m}}(\mu^2 - 1)^l \qquad ...(5.62)$$

This is the solution to θ equation.

Spherical Harmonics

Since the solution to both the θ equation and the solution to the Φ equation contain trigonometric functions so to determine the angular character of the electron wave function. We can use the total wave function conveniently if it is separated into a *raaial* portion and an *angular* portion in such a way that

$$\Psi_{(r\,\theta_{\varphi})} = R_{n'}(r)\, Y_{m'}(\theta, \varphi)$$

The term $Y_{l,m}(\theta_\psi)$ is known as the spherical harmonics, and is given by

$$Y_{l,m}(\theta,_\varphi) = \theta_{l,m}(\theta)\,\Phi. \qquad ...(5.63)$$

This portion to the total wave function will be of major concern in the treatment of directional bonding.

The Radial Equation

The radial Equation is

$$\frac{1}{r^2}\frac{d}{dr}\left(r^2\frac{dR_{(r)}}{dr}\right) - \frac{\beta R_{(r)}}{r^2} + \frac{8\pi^2\mu}{h^2}\left[E + \frac{Ze^2}{r}\right] = R_{(r)} = 0 \qquad ...(5.64)$$

This equation like the θ equation, can be put into such a form that has long been known to the mathematicians. This particular equation is the *Laguerre equation*, and its normalised solution is given by

$$R_{n,l}(r) = \sqrt{\left(\frac{2Z}{na_0}\right)^3 \frac{(n-l-1)!}{2n[(n+l)!]^3}}\; e^{-\rho/2}\rho^l L_{n+l}^{2l+1}(\rho) \qquad ...(65)$$

where $\rho = (2Z/na_0)\,r$, $a_0 = h^2/4\pi^2\mu e^2$, and $L_{n+l}^{2l+1}(\rho)$ represents the associated Laguerre polynomial. Like the θ equation, the solution to the radial equation is also complex. But it is possible to make some observation from the solution. It may be noted that a new parameter the quantum number n, has been added. Although we have not solved the radial equation, its solution is analogous to that of the θ equation. Thus, the new quantum number n is restricted to take on the integral values 1, 2, 3,.....n. Both the relation of n to the radial wave function, which is a measure of the position of the electron with respect to the nucleus, and its similar restrictions shows that n is the quantum mechanical analog of the principal quantum number of the Bohr Theory.

Using the same approach as that applied to the magnetic quantum number, a new restriction can be seen for the quantum number l. It is clear from the normalizing factor of the solution to the radial equation that the term $(n - l - 1)$ demands that the maximum value of l be $(n - 1)$. If l is allowed a value greater than this, the factorial of a negative number would result. Thus, the quantum number is restricted to the values l = 0, 1, 2, ..., $(n - 1)$.

Solution of the Radial Equation

The radial equation can be written as

$$\frac{1}{R}\frac{d}{dr}\left(r^2\frac{dR}{dr}\right)+\frac{8\pi^2 m\, r^2}{h^2}\left(E+\frac{Ze^2}{r}\right)=\beta=l\,(l+1) \qquad ...(5.66)$$

Multiplying it by R/r^2, we get

$$\frac{1}{r^2}\frac{d}{dr}\left(r^2\frac{dR}{dr}\right)-\frac{l\,(l+1)}{r^2}R+\frac{8\pi^2 mR}{h^2}E$$

$$+\frac{8\pi^2 mZe^2 R}{h^2 r}=0 \qquad ...(5.67)$$

Now the total energy E for a bound electron is negative. Thus

let $$\frac{8\pi^2 mE}{h^2}=-\alpha^2 \qquad ...(5.68)$$

Assuming the equation to be valid for large and small values of *r* we have

$$\frac{1}{r^2}\frac{d}{dr}\left(r^2\frac{dR}{dr}\right)=\frac{1}{r^2}\cdot r^2\frac{d^2R}{dr^2}+\frac{2}{r}\frac{dR}{dr}$$

$$=\frac{d^2R}{dr^2}+\frac{2}{r}\frac{dR}{dr}. \qquad ...(5.69)$$

Hence, equation (5.65) can be put as

$$\frac{d^2R}{dr^2}+\frac{2}{r}\frac{dR}{dr}-\frac{l\,(l+1)}{r^2}R+\frac{8\pi^2 mR}{h^2}E$$

$$+\frac{8\pi^2 mZe^2 R}{h^2 r}=0 \qquad ...(5.70)$$

For an asymptotic solution when *r* is very large, the terms

$$\frac{2}{r}\frac{dR}{dr},\ \frac{l\,(l+1)}{r^2} \text{ and } \frac{8\pi^2 mZe^2}{h^2 r}$$

in equation (5.68) can be neglected. Hence it becomes

$$\frac{d^2R}{dr^2}+\frac{8\pi^2 mE}{h^2}R=0 \text{ (by 50.13)} \qquad ...(5.71)$$

or $$\frac{d^2R}{dr^2}-\alpha^2 R=0, \qquad ...(5.72)$$

where $$R=Ae^{\pm\,\alpha r} \qquad ...(5.73)$$

Now the solution $R = Ae^{+\alpha r}$ is not admissible because R (*i.e.* ψ) has to be finite but this solution makes it infinite for large values of r. Hence $R = Ae^{-\alpha r}$.

When r is very small, of the order of zero, equation (68) because

$$\frac{dR^2}{dr^2} + \frac{2}{r}\frac{dR}{dr} - \frac{l(l-1)}{r^2} R = 0 \qquad ...(5.74)$$

(since $f/(r)$ is small, $\alpha^2 R$ can be neglected).

Let $R = a_0 + a_1 r + 3a_2 r^2 + ...$ be the required solution.

then $$\frac{dR}{dr} = a_1 + 2a_2 r + 3a_2 r^2 + ...,$$

and $$\frac{dR^2}{dr^2} = 2.1a_2 + 3.2a_3\, r + ...$$

Equation (74) then becomes

$$(1.2a_2 + 2.3a_2 r + ...) + \left(\frac{2a_1}{r} + 2.2a_2 + 2.3a_2\, r + 1...\right)$$

$$-\frac{l(l+1)}{r^2} a_0 - \frac{l(l+1)}{r} a_1 - l(l+1)\, a_2 - l(l+1)\, a_3 r - ... = 0 \qquad ...(5.75)$$

In this identity the coefficients of like powers of r must be equated to zero. So considering the coefficients of r we get

$$3.2a_2 + 2.3a_3 - l(l+1)\, a_2 = 0 \qquad ...(5.76)$$

or $$3.4a_3 - l(l+1)\, a_2 = 0 \text{ or } 3(3+1)\, a_3 - l(l+1)\, a_2 = 0$$

Considering the general term and calling it v, we get

$$[(v+1)\, v - l(l+1)]\, av = 0. \qquad ...(5.77)$$

As $a, \neq 0$, because otherwise the series will vanish.

$$(v+1)\, v - l(l+1) = 0$$

or $$v = l.$$

If $v < l$ or $> l$, $v(v+1) - l(l+1) \neq 0$ and hence $a, = 0$ *i.e.*, the series disappears. The required solution for small values of r is at r^l. Hence the general solution is series of the form :

$$R = Ae^{-\alpha r} r^l (b_0 + b_1 r + b_2 r^2 + ...). \qquad ...(5.78)$$

The solution of the radial equation is given below :

$$\frac{1}{r^2}\frac{d}{dr}\left(r^2\frac{dR}{dr}\right) - \frac{l(l+1)R}{r^2} + \frac{8\pi^2 mE}{h^2} \cdot R + \frac{8\pi^2 me^2 Z}{h^2 r} R = 0$$

(Radial Equation)

By considering E as negative and writing

$$\frac{8\pi^2 mE}{h^2} = -\alpha^2$$

Introducing a dimensionless independent variable

$$\rho = 2\,\alpha r,$$

radial equation on multiplication by r^2 becomes

$$\frac{d}{dr}\left(r^2\frac{dR}{dr}\right) - l(l+1)R - \alpha^{2/2}R + \frac{8\pi^2 mZe^2 r}{h^2}R = 0 \quad ...(5.79)$$

or
$$\frac{d}{d\rho}\left(\rho^2\frac{dR}{d\rho}\right) - l(l+1)R - \frac{\rho^2}{R}R + \frac{8\pi^2 mZe^2}{h^2}\cdot\frac{\rho}{2\alpha}R = 0$$

or
$$\frac{d}{d\rho}\left(\rho^2\frac{dR}{d\rho}\right) - l(l+1)R - \frac{\rho^2 R}{4} + \lambda\rho R = 0$$

where
$$\lambda = \frac{4\pi^2 mZe^2}{h^2\,\alpha}. \quad ...(5.80)$$

This equation can also be solved for large and small values of r.

(*i*) For large values of r, the solution is, $R = e^{-\rho/2}$.

(*ii*) For small values of r the solution is

$$R = Ar = A'(2\alpha r)^l$$
$$= A'\rho^l, \text{ where } A = A'(2\alpha)^l = \text{constant.}$$

Let the general solution be

$$R = e^{-\rho/2}\rho^l(a_0 + a_1\rho + a_2\rho^2 + \ldots.)$$

Then
$$\frac{dR}{d\rho} = \frac{1}{2}e^{-\rho/2}p^l(a_0 + a_1\rho + a^2\rho^2 + \ldots.).$$
$$+ e^{-\rho/2}l_\rho^{l-1}(a_0 + a_1\rho + a_2\rho^2 + \ldots.)$$
$$+ e^{-\rho/2}\rho^l(a_1 + 2a_2\rho + 3a_2\rho^2 + \ldots.)$$

$\therefore$
$$\rho^2\frac{dR}{d\rho} = -\tfrac{1}{2}e^{-\rho/2}\rho^{l+2}(a_0 + a_1\rho + a_2\rho^2 + \ldots.)$$
$$+ l_\rho^{l+1}e^{-\rho/2}(a_0 + a_1\rho + a_2\rho^2 + \ldots.)$$
$$+ e^{-\rho/2}\rho^{l+2}(a_0 + 2a_3\rho + 3a_2\rho^2 + \ldots.)$$
$$\frac{d}{d\rho}\left(\rho^2\frac{dR}{d\rho}\right) - l(l+1)R - \frac{\rho^2}{4}R + \lambda\rho R$$

$$= [\frac{1}{4}\rho^{l+2} e^{-\rho/2} (a_0 + a_1\rho + a_2\rho^2 + \ldots)$$

$$- \frac{1}{2}(l+2)\rho^{l+1} e^{-\rho/2} (a_0 + a_1\rho + a_2\rho^2 + \ldots)$$

$$- \frac{1}{2} e^{-\rho/2} \rho^{l+2} (a_1 + 2a_1\rho + 3a_3\rho^2 + \ldots)]$$

$$+ [-\frac{1}{2} e^{-\rho/2} l\rho^{l+1} (a_0 + a_1\rho + a_2\rho^2 + \ldots)$$

$$+ l(l+1)\rho^l e^{-\rho/2} (a_0 + a_1\rho + a_2\rho^2 + \ldots)$$

$$+ l\rho^{l+1} e^{-\rho/2} (a_1 + 2a_2\rho + 3a_3\rho^2 + \ldots)$$

$$+ [-\frac{1}{2} e^{-\rho/2} \rho^{l+2} (a_1 + 2a_2\rho + 3a_3\rho^2 + \ldots)$$

$$+ (l+2)\rho^{l+1} e^{-\rho/2} (a_1 + 2a_2\rho + 3a_3\rho^2 + \ldots)$$

$$+ e^{-\rho/2} \rho^{l+2} (2a_2 + 3.2a_2\rho + \ldots)$$

$$+ [-l(l+1)\rho^l e^{-\rho/2} (a_0 + a_1\rho + a_2\rho^2 + \ldots)]$$

$$- \left[\frac{\rho^{l+2}}{4} e^{-\rho 2} (a_0 + a_1\rho + a_2\rho^2 + \ldots]\right.$$

$$+ [\lambda - e^{-\rho/2} \rho^{l+1} (a_0 + a_1\rho + a_2\rho^2] = 0 \qquad \ldots(5.81)$$

In this identity the sum of the coefficients of the like powers of ρ must be equated to zero. Equating to zero the coefficient of $e^{-\rho/2}\rho^{l+2}$ we get

$$-\frac{1}{2}(l+2)a_1 - \frac{1}{2}a_1 - \frac{1}{2}la_1 + 2la_2 - \frac{1}{2}a_1 + (l+2).2a_2 + 2a_2 + \lambda a_1 = 0$$

or $$(\lambda - l - 2)a_1 + 2(2l + 2 + 1)a_2 = 0$$

or $$(\lambda - l - 1 - 1)a_1 + (1+1)(1 + 2l + 2)a_2 = 0$$

or $$(\lambda - l - k - 1)a_2 + (k+1)(k + 2l + 2)a_{2+1} = 0$$

or $$+1 = \frac{(k + l + 1 - \lambda)}{(k+1)(k+2l+2)} a_k \qquad \ldots(5.82)$$

This series is normally unacceptable because it is a divergent series, hence the wave function does not tend to zero as the series tends to infinity. Consequently we have to assume that the series is a polynomial of degree k, such that the coefficients of terms after the kth terms of the zero.

Hence, if $a_{2+1} = 0$, it is essential that

$$k + l + 1 - \lambda = 0$$

$$\lambda = k + l + 1 = n, \text{ where } n = k + l + 1. \quad ...(5.83)$$

The above series, in which each term is related to each preceding term *i. e.* a_2, to a_1, a_3 to a_2, etc. is known as **Laguerre's polynomial.** *The series cannot be broken up into odd series and even series.*

Total Wave Function of the Hydrogen-like Atoms

The complete wave functions of the hydrogen-like atoms in spherical polar coordinates are given by

$$\psi (n, l, m) = R_{n, l(r)} \, Y_{l, \pm m} (\theta, \phi) \quad ...(5.84)$$

where $Y_{l, \pm m}$ are the normalised spherical harmonics. It may be noted that the quantum number n determined the energy of the orbital, l, the angular momentum and m, the z-component of the angular momentum. The orbital with l= 0, 1, 2, 3 etc. are known as s, p, d, f orbitals respectively. An orbital with $n = 1, l = 0$ is known as 2 1s orbital, with $n = 3, l = 1$ is known as 3p orbital, etc. There are 3p orbitals with $m = 0, + 1, - 1$, and they are designated as p_x, p_y and p_z or as p_0, p_1, and p_{-1}. The angular factors associated with these orbital are as under

$$p_0 \sim \cos \theta \quad ...(5.85)$$

$$p_{+1} \sim \sin \theta \exp (i \psi)$$

$$p_{-1} \sim \sin \theta \exp (- i \phi)$$

These are complex functions so, we cannot draw their picture in real space. However, we can get functions that are equally good solutions. Taking the following combinations of spherical harmonics with the same l values. We have

$$-\frac{1}{\sqrt{2}} (Y_{15+1} + Y_{15-1}) = \frac{\sqrt{3}}{2\sqrt{\pi}} \sin \theta \cos \phi$$

$$- i/\sqrt{2} \, (Y_{15+1} - Y_{15-1}) = \frac{\sqrt{3}}{2\sqrt{\pi}} \sin \theta \sin \phi \quad ...(5.86)$$

Because sin θ cos φ and sin θ sin φ are respectively the angular dependence of x and y components of r the above two combinations are respectively known as p_x and p_y, atomic orbitals. The total wave functions for different components characterised by $n = 2$, $l = 1$ are shown in Table 5.2.

TABLE 5.1

n	l	m	*Wave Functions*
2	1	0	$\psi_2 p_z = \frac{1}{4\sqrt{2\pi}}\left(\frac{Z}{a_0}\right)^{3/2} r\cos\theta \exp\left(-\frac{Zr}{2a_0}\right)$
2	1	± 1	$\psi_2 p_x = \frac{1}{4\sqrt{2\pi}}\left(\frac{Z}{a_0}\right)^{3/2} r\sin\theta\sin\phi \exp\left(-\frac{Zr}{2a_0}\right)$
			$\psi_2 p_x = \frac{1}{4\sqrt{2\pi}}\left(\frac{Z}{a_0}\right)^{3/2} r\sin\theta\cos\phi \exp\left(-\frac{Zr}{2a_0}\right)$

Quantum Numbers

We have already seen that $n = k + l + 1$

where k = an integer

and l = an integer

Hence, n is also an integer. It involves k, which can take up values 0, 1, 2, 3 etc.

(i) When $n = 1$, $k = 0$ and $l = 0$

(ii) When $n = 2$, $k = 0$ and $l = 0$

or $k = l$ and $l = 0$

Similarly l can have n values and it can also have a value of 0. So the values of 'l' are

$$l = 0, 1, 2, 3, \ldots (n - 1);$$

'l' is known the *azimuthal quantum number.*

'm' which is involved in l is known as the *magnetic quantum number.* It can have values

$$m = 0, \pm 1, \pm 2 + \ldots \pm l.$$

Hence, m can have $(2l + 1)$ values for any given value of l.

We have $$n^2 = \lambda = \frac{4\pi^2 mZe^2}{h^2\alpha}$$

or $$n = \frac{16\pi^4 m^2 Z^2 e^4}{h^4\alpha^2}$$

But $$a^2 = -\frac{8\pi^2 mE}{h^2}$$

Therefore $$n^2 = -\frac{16\pi^4 m^2 Z^2 e^4 h^2}{h^4 \cdot 8\pi^2 mrE}$$

or $$E = -\frac{2\pi^2 mZ^2 e^4}{n^2 h^2} \cdot \quad ...(5.87)$$

This quantity is identical with Bohr's equation for the total energy of the hydrogenic atom. This indicates that the energy of such an atom in the state represented the quantum number n, l, m does not depend on the individual values of these quantities but only on the total *principal quantum number, n.*

Quantum States

From the solution of the total wave equation, we have obtained three quantum numbers. These numbers had been postulated in the Bohr-Sommerfeld theory, and this was considered a fundamental weakness of the theory. It is, thus satisfying to see that they occur as the result of a few basis postulates in the wave mechanical treatment. However, it is unfortunate that they can no longer be considered of in the pictorial manner of the Bohr theory. We now have one quantum number for each degree of freedom, but the ideas of processing orbits is no longer valid.

The quantum number with their permitted values can be summarized as under.

Radial quantum number $n = 1, 2, 3...,$
Azimuthal quantum number $l = 0, 1, 2.., n - 1$ (inclusive)
Magnetic quantum number $m = 0, \pm 1, \pm 2,..., \pm l$ (inclusive)

According to these restrictions, there are only certain values of the quantum number l which are permissible for a given value of n. The maximum value of l is to be $(n - 1)$. For example, when $n = 4$, l can be any integer up to and including $l = 3$, but no greater. This is shown in Table 5.2, where the allowed values of l are shown for the first four radial shell.

TABLE 5.2
Quantum Numbers with Allowed Values for First Four Radial Shells

Value of n	1	2	3	4
Allowed Values of l	0	0, 1	> 0, 1, 2	0, 1, 2, 3

It may be noted that $l = 0$ occurs for every value of n, $l = 1$ occurs for every value of n greater than $n = l$, an so on. These values of the quantum number l are important both in the geometry and the energy states of the atom. Because of this importance, they are given the special designations as under :

$l = 0$	*s* state
$l = 1$	*p* state
$l = 2$	*d* state
$l = 3$	*f* state

It is common to speak of the electron state that the particular value of the l quantum number.

For the first radial shell, the value of the radial quantum number is $n = 1$, and the l quantum number can have only the valve $l = 0$. This state is generally represented by (1*s*) where the one represents the value of the n quantum number and *s* represents $l = 0$. For $n = 2$, the azimuthal quantum number can have the values $l = 0$ and $l = 1$. This gives the two states (2*s*) and (2*p*), respectively. For the case of $n = 3$, it can be seen from the allowed values of the l quantum number that there can be the three states (3*s*), (3*p*), and (3*d*) for $l = 0$, $l = 1$, and $l = 2$, respectively. Finally, for the fourth shell for which $n = 4$, it is seen that there can be the 4 states (4*s*), (4*p*), (4*d*) and (4*f*). These states determined the possible energies of the electrons, and if the l quantum number contributes to the energy as does the n quantum number, each state will represent a different energy.

Energy States of the Hydrogen Atom

In the wave equation of the hydrogen-like atom, the energy term are found only in the radial equation; and it is from its solution that the energy states of the hydrogen atom are obtained. As the potential energy of the electron with respect to the nucleus is taken as zero when the two particles are infinitely far apart, the potential energy becomes increasingly negative as the electron approaches the nucleus. Therefore the energy of the electron is always the negative energy. If only these negative stages are considered, it can be shown that the permitted values of the energy of the electron are given by

$$E_n = -\frac{2\pi^2 \mu Z^2 e^4}{n^2 h^2}, n = 1, 2, 3, \ldots \quad \text{...(5.88)}$$

This expression for the energy of the electron as determined by wave mechanics is found to be identical with that of the Bohr theory.

The quantum number, n, which appears in the wave mechanical treatment is related to the principal quantum number of the Bohr theory and can therefore be associated with the electron shell. Thus $n = 1$ corresponds to the first or *K* shell, and for $n = 2$ we have the L shell etc.

It may be noted that the energy of the electron in the hydrogen like atom is determined only by the *n* quantum number according to our wave equation and the resulting energy levels must be degenerate. The ground state of the hydrogen atom will be the (1) state, and the subsequent order of energy levels is given by

$$1s < 2s = 2p < 3s = 3p = 3d < 4s = 4p = 4d = 4f < 5s.....$$

Since the *l*, *m*, and *m*, quantum numbers contribute nothing to the energy of the electron state, all of the possible energy states in a given radial shell have equal energy. Thus not only single spectral lines such as those predicted by Bohr will be observed, but a fine structure exists in the hydrogen spectrum. This let to the development of the Bohr-Sommerfeld theory of the hydrogen atom.

Actually the sequence of levels obtained is the result of calculations based upon a nonrelativistic hydrogen-like atom in the absence of external electric or magnetic fields. By using a relativistic form of the wave equation, the orbital degeneracy can be removed, thereby leading to the experimentally observed fine structure. Unfortunately, because of the extreme complexity of the mathematics, a relativistic treatment is not practical. For more complex atoms we well see that the orbital degeneracy can be removed by considering the effect of the electron repulsion.

Wave Functions of the Hydrogen Atom

We know that the wave function depends on the relation

$\psi = R \cdot \Theta \Phi$, where R depends on n

Θ depends on l

and Φ depends on m.

Therefore ψ depends on *n*, *l* and *m* and it is designated as $\psi(n, l, m)$. The total ψ can be written as under

$$\psi(n, l, m) = Ae^{-\rho/2} \cdot \rho^l (\Sigma a\rho^k).$$

$$\sin^m \theta \frac{d^{l+m}}{d\mu^{l+m}} (\mu^2 - 1)^l e^{\pm im\phi} \quad ...(5.89)$$

in the un-normalised state.

Wave Functions of Different Electrons

Making use of equation (5.89) it is possible to arrive at the wave functions of electrons in different orbits.

(i) When $n = 1$, $l = 0$ and $m = 0$, by substituting for l and m, we get

$$\psi_{100} = Ae^{-\rho/2} \cdot a_0 \cdot 1 = \text{const. } e^{-\rho/2} \qquad \text{...(5.90)}$$

(ii) When $n = 2$,

l may be 0, 1

and m ...... 0. ± 1.

$$(a)\ \psi_{200} = \text{const, } e^{-\rho/2}(a_0 + a_1\rho)\ 1.1 = Ce^{-\rho/2} + D\rho\, e^{-\rho/2}$$

$$= e^{-\rho/2}(C + D\rho) \qquad \text{...(5.91)}$$

$$(b)\ \psi_{210} = A \cdot e^{-\rho/2} \rho \cdot a_0\, 2\cos\theta = \text{const. } \rho\cos\theta\, e^{-\rho/2} \qquad \text{...(5.92)}$$

$$\psi_{21,\pm 1} = Ae^{-\rho/2} a_0\rho \sin\theta\, 2e^{\pm i\phi} = \text{const. } e^{-\rho/2}$$

$$\rho\sin\theta\, e^{\pm i\phi} \qquad \text{...(5.93)}$$

$$\psi_{211} = A \cdot e^{-\rho/2} a_0\, \rho\sin\theta\, e^{-i\phi} = \text{const. } e^{-\rho/2}$$

$$\rho\sin\theta\, e^{-i\phi} \qquad \text{...(5.94)}$$

$$\psi_{211} = Ae^{-\rho/2} a_0\, \rho\sin\theta\, e^{i\phi}$$

$$= Ae^{-\rho/2} a_0\, \rho\sin\theta\,(\cos\phi + i\sin\phi)$$

$$\psi_{211} = Ae^{-\rho/2} a_0\, \rho\sin\theta\cos\phi = \text{const. } e^{-\rho/2}\rho\sin\theta\cos\phi \qquad \text{...(5.95)}$$

$$\text{and also=const. } e^{-\rho/2}\rho\sin\theta\sin\phi \qquad \text{...(5.96)}$$

The Electron Spin

In addition to the orbital fine structure that can be explained in terms of the l quantum number, it was experimentally found that a doublet structure also exists. Spectral lines which had once thought as single lines actually came out to be two lines very close together. This doublet character could not be explained by the Bohr-Sommer feld model. In 1925, Uhlenbeck and Goudsmit tried to explain it when they proposed that the electron has, in addition to its orbital motion, an angular momentum of rotation about its own axis and a corresponding *magnetic moment*. This gave rise to a new quantum number called *the spin quantum number*. The magnitude of this spin angular momentum is found to be $\pm\frac{1}{2}$ in units of $h/2\pi$. The plus or minus values of the spin arise from the direction of the spin. For example, if the electron spin is clockwise it interacts with the orbital magnetic moment of the electron to give a different energy than if it is counterclockwise. The energy difference resulting from these opposite spins is small, but it is sufficient to give rise to the observed of doublet structure. Actually, there are many difficulties that arise from the concept of a physically spinning electron. However, the agreement between theory and experiment is sufficiently good to retain the basic idea.

It was only in 1928 that a quantum mechanical answer was found to the problem of electron spin. The wave equation as developed by Schroedinger was nonrelativistic. In an attempt to bring wave mechanics into harmony with relativity theory. Dirac succeeded in developing a wave equation which led naturally to the proper spin angular momentum for the electron. According to the Dirac theory the electron has the same angular momentum and magnetic moment as the spinning electron of Uhlenbeck and Goudsmit. Yet, the quantum mechanical properties of the electron spin are the result of a consistent mathematical treatment and do not offer the problems which arises from the physical picture of an electron spinning on its own axis.

The Term ρ and Bohr's Orbit

In our previous discussion we have assumed $\rho = 2\alpha r$ and we now find that this parameter has been quite useful to express the wave functions not only of the hydrogenic atom but also of the electrons in different states. Now we will try to explore the physical significance of ρ.

$$\rho = 2\alpha r$$

or $$\rho^2 = 4x^2 r^2 = 4\left(-\frac{8\pi^2 m}{h^2}E\right)r^2$$

$$= \left(-\frac{32\pi^2 m}{h^2}\right)\left(\frac{2\pi^2 mZ^2e^4}{n^2h^2}\right)r^2$$

$$= \frac{64\pi^4 m^2 Z^2 e^4 r^2}{n^2 h^2}$$

or $$\rho = \frac{8\pi^2 mZe^2}{nh^2}r. \qquad ...(5.97)$$

Now if a_0 be the first Bohr orbit then

$$a_0 = \frac{h^2}{4\pi^2 me^2}$$

and $$\frac{1}{a_0} = \frac{4\pi^2 me^2}{h^2} \quad ...(5.98)$$

and $$\rho = \frac{8\pi^2 mZe^2 r}{nh^2} = \frac{2Zr}{na_0} \quad ...(5.99)$$

So the parameter ρ is connected with the first orbit of Bohr by the equation (5.99).

The Orbitals

(i) The s-orbitals **: A wave function of an electron is known as an *orbital or state*. All those orbitals for which n has integral values but l and m are zero are designated as the *s-orbital*.**

Thus when $n = 1$ and $l = m = 0$, we have 1 s orbital

$n = 2$and $l = m = 0$ we have 2s orbital

and $$\psi_{100} = Ae^{-\rho/2} \text{ from equation (5.90)}$$

$$= Ae^{-Zr/na_0} \text{ by (5.99)}$$

$$= Ae^{-r/a_0} \text{ since } n = 1. \quad ...(5.100)$$

$$\psi_{200} = (C + D_\rho)\, e^{-\rho/2} \text{ from (5.91)}$$

$$= \left(C + D.\frac{2Zr}{2a_0}\right) e^{Z - r/na_0}$$

$$= \left(C + D\frac{Zr}{a_0}\right) e^{-Zr/2a_0} \quad ...(5.101)$$

Thus we find that the 1s state and 2s state are independent of θ and φ, so these orbitals will be spherically symmetrical and the projections of this orbit on a plane will be circular one.

Probability Density in 1*s* Atomic Orbital

If we examine the 1 s orbital of the hydrogen atom (Z=1) the according to the wave mechanical theory we have

$$\psi_{1s} = \left[\frac{1}{\pi a_0^3}\right]^{1/2} \exp\left(-\frac{r}{a_0}\right)$$

where r is the distance from the nucleus. The probability density for an electron in this orbital is ψ_{1s}^2 and is given by

$$\psi_{1s} = \frac{1}{\pi a_0^2} \exp\left(-\frac{2r}{a_0}\right)$$

should be noted ψ_{1s}^{2} is independent of θ and ϕ and so the electron distribution is 1s orbital is spherically symmetrical. The plot ψ_1 against r/a_0 is shown in Fig. 5.3. The electron distribution in an orbital can also be described by determining the probability of an electron being found in a spherical shell of thickness at a distance r from the nucleus. It is given by

$$P(r)\,dr = 4\pi r^2 \psi_{1s}^{2}\,dr$$

$$= 4\pi r \frac{1}{\pi a_0^3} \exp\left(-\frac{2r}{a_0}\right) dr$$

We plot $P(r)$ against (r/a_0) in Fig. 5.3. The maximum is found at value of r which is found from the equation

$$\frac{\partial P(r)}{\partial r} = \left[-\frac{\partial r^2}{a_0} + 2r\right] \exp\left(\frac{-2r^2}{a_0}\right) = 0$$

$$r = a_0$$

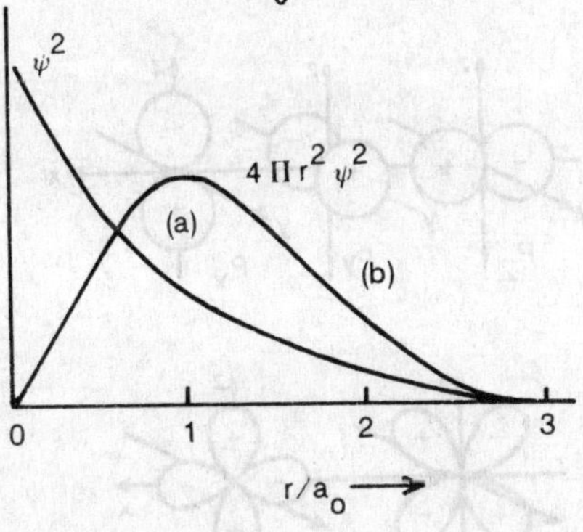

Fig. 5.3. *Probability densities for the 1s hydrogen atomic orbitals.*

Hence the maximum radial density occurs at a_0, therefore the most probable distance of electron from the nucleus is precisely what has been predicted from the Bohr theory.

It may appear very puzzling to note that the electron density is maximum right on the nucleus whereas the curve shows that electron is most probably found at a distance a_0 from the nucleus. This apparent inconsistency can be understood from the following illustration. Consider a number of concentric rings of same thickness but of different radii and made of different materials in which the density of the material decreases as the radius increases *i.e.*, the ring closest to the centre is made of the densest material but does not have the highest mass, because of the smallest volume. On the other hand, a ring of a very large radius may not have the highest mass due to the very small density of its material. But, for a certain intermediate value of the radius the ring should have the largest mass although the density of the material is maximum for the ring does not to the centre.

The p-orbitals

For all those states for which $l = 1$, the designation is p-orbitals. These p-orbitals can have three different orientations depending on the permitted values of m (0 or $\pm$ 1),

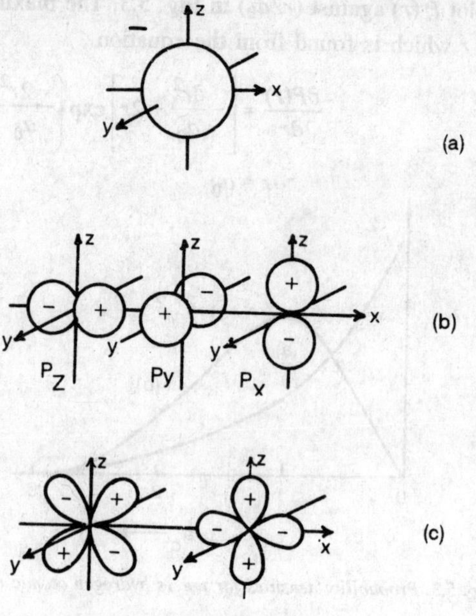

Fig. 5.4.

We have seen from equations (5.92), (5.95) and (5.96) that

$$\left.\begin{aligned} \psi_{210} &= \text{const.} f(r) \cos\theta \\ \psi_{211} &= \text{const.} f(r) \sin\theta \cos\phi \\ \text{and} \quad &= \text{const.} f(r) \sin\theta \sin\phi \end{aligned}\right\} \quad \text{...(5.102)}$$

Comparing these with the relations between the cartesian and polar co-ordinates

$$\left.\begin{array}{llr} z \text{ component of } r & = & r\cos\theta \\ x \quad \ldots & & r = r\sin\theta\cos\phi \\ y \quad \ldots & & r = r\sin\theta\sin\phi, \end{array}\right\} \qquad \ldots(5.103)$$

We find that ψ_{210} is the z-component. This is known as p_2 orbital.

$\psi_{211} = f(r)\sin\theta\cos\phi$ is known as the p_x orbital ...(5.103)

$\psi_{211} = f(r)\sin\theta\sin\phi$ ...... p_y orbital

Since $n = 2$, these are commonly known as $2p_x$, $2p_y$ and $2p_x$ orbitals.

The plot of p_z function gives the cosine curve ($\cos\theta$) and if we plot for a complete rotation of ϕ it takes the shape of a dumb-bell as shown in the Fig. 5.2 (*b*). It shows p-electrons with $m = 0$. The p_z electron has no probability along x and y direction, similarly p_x and p_y have no probability along y and z, and z and x respectively. They are shown in the same fig. side by side with p_z. The actual disposition of electron clouds for the different orbital is shown in next chapter.

The d-orbitals

For *d*-orbitals we have to take up those states where $n = 3$ and $l = 2$. The state for which for $n = 3$ and $l = 1$ are called the $3p$ orbitals and their wave functions show the same characteristics as $2p$. Only the curves are more elongated.

For $n = 3$ and $l = 2$, we can derive that

$$\psi_{3,2,0} = f(r)(3\cos^2\theta - 1)$$

$$\psi_{3,2\pm1} = f(r)\sin\theta\cos\theta(\cos\phi \pm i\sin\phi). \qquad \ldots(5.104)$$

and $\psi_{3,2\pm2} = f(r)\sin^2\theta(\cos 2\phi \pm i\sin 2\phi)$.

As regards the shape of the *d*-orbitals we find that the curve expressed by $(3\cos^2\theta - 1)$ is a loop and the distribution of the wave function has been shown in the figure for the *d*-states [Fig. 7.2. (*c*)] when $m = 0$. Pauling designated this state d_z state while Eyring prefers to designate this as d_z^2 state.

For $\psi_{3,2,\pm1}$ and $\psi_{3,2,\pm2}$ states the distributions are similarly shown. From the figure we find that the distribution is more symmetric in the *d*-state than in the *p*-state.

Shape of Atomic Orbitals

We have already studied the behaviour of 1*s* atomic orbital and the probability functions in terms or *r*. It is found that the wave functions for 1*s* orbital and for any other *s* orbitals having $l = 0$ are spherically symmetrical. So it is possible to draw a spherical boundary surface such that nearly all the electron density (say 99%) can be found inside the surface. In a plane, Fig. 5.4. (*a*) can represent the shape of *s*-type atomic orbitals. But it is to be kept in mind that the orbitals increase in size as *n* increases.

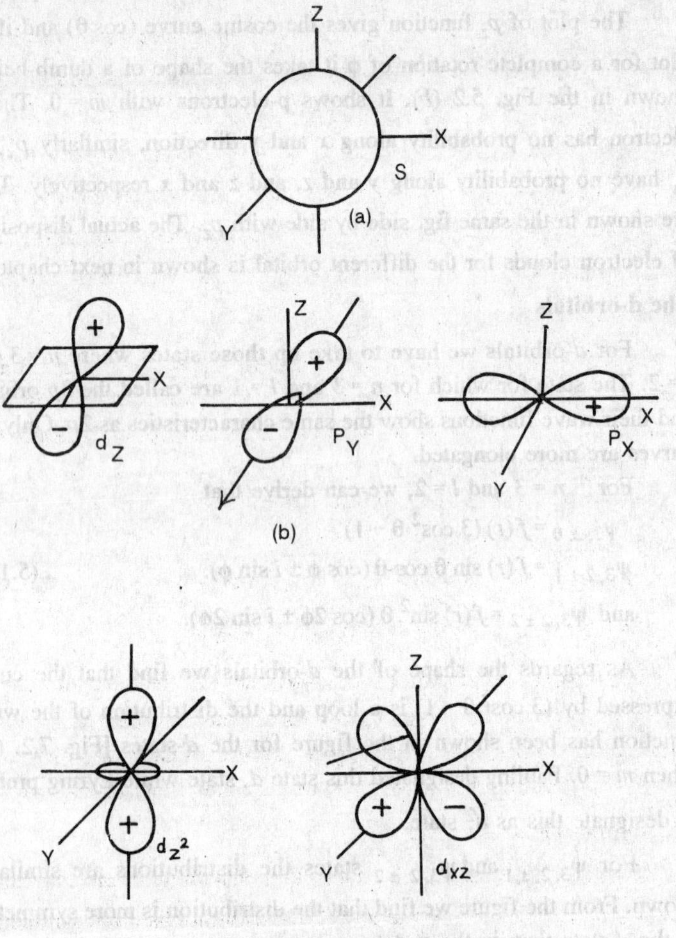

Fig. 5.5. *Shape of atomic orbitals*

The wave functions for the *p* type of atomic orbitals for which (l = 1) has been studied and it has been found that the angular functions have only the quantum numbers l and m whereas the *r*-functions depend on n and l. Because shapes of orbitals are controlled by the angular functions so it depends on l and m. Considering $2p_z$ atomic orbital whose wave function can be written as

$$\psi\, 2p_z = R_{2,1}(r) \cos\theta \quad \text{...(5.105)}$$

Thus, we find that the probability of finding an electron at $\theta = \pi/2$, *i.e.* in the X-Y plane, is zero. Hence the X-Y plane is called a nodal plane. The charge is therefore distributed above and below this plane and exhibits a rotational symmetry around the Z-axis because ψ_{2pz} is independent of ϕ. This atomic orbital has the shape of a dumbbell with positive and negative lobes [5.4 (*b*)]. The other two *p*-orbitals *viz.* ψ_{2px} and ψ_{2py}, have respectively Y-Z and Z-X as their nodal planes. These atomic orbitals have the regions separated by the nodal plane and have opposite signs. However, no physical significance can be attributed to signs of ψ. The physically significant ψ^2 or $\psi^* \cdot \psi$ is always positive.

In all there are five *d* orbitals (l - 2) of which two are shown in Fig. 5.4 (*c*). In many respects a *d* orbital is like the superposition of two *p*-orbital because that are four similar regions of alternating sign which are separated by 2 nodal planes. The d_z^2 orbital has two large pear-shaped lobes pointing along the Z-axis with a central collar. Similarly, the *f*-type and *g*-type atomic orbitals can be visualized but they are more complicated.

Thus, we find that only *s*-type of orbitals are spherical and have finite density on the nucleus. This is important in the coupling between the magnetic moment because of electron and that because of nucleus in magnetic resonance spectroscopy. All other types of orbitals have nodes on the nucleus and have special directions in space where the functions have a large value and these special directions play an important role in the theory of valence.

Physical Interpretations of the Hydrogenic Orbitals

We can prove that the operator $\hat{L}^2$ commutes with the atomic hamiltonian. Therefore, the hydrogenic wave functions are the eigen-function of the operator for the square of the angular momentum of the electron about an axis passing through the nucleus. If $\hat{L}^2$ has the eigenvalues $l(l+1)/(h/2\pi)^2$, the modulus of the orbital angular momentum vector has the value $[l(l+1)]^{1/2}\,h/2\pi$ or $\sqrt{l(l+1)}$ in atomic units, where $h/2\pi = 1$.

If an electron of mass m, moves about the nucleus in some definite orbit with a velocity v, then the centrifugal force which tends to keep the electron away from the nucleus is given by

$$F = \frac{m_2 v^2}{r} = \frac{m_e^2 v^2 r^2}{m_e r^2} = \frac{L^2}{m_e r^2} \quad \text{...(5.106)}$$

where r is the distance between the electron and the nucleus.

For an electron, making use of $m_e = 1$, and $L^2 = l(l+1)$, F is given by

$$F = \frac{l(l+1)}{r^2} \quad \text{... (5.107)}$$

Hence, the radial motion of an electron in an atom is controlled not only by Coulombic force $-Z/r^2$ but also by a centrifugal force which is given by Eq (5.107). The Coulombic force tends to bring electron closer to the nucleus but the centrifugal force tries to keep the two particles apart for a stable atom.

$$\frac{l(l+1)}{r^2} = \frac{Z}{r^2} = 0$$

$$r = \frac{l(l+1)}{Z} \quad \text{...(5.108)}$$

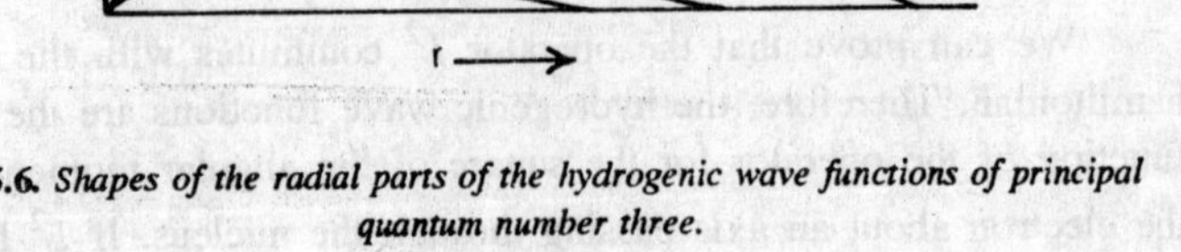

Fig. 5.6. *Shapes of the radial parts of the hydrogenic wave functions of principal quantum number three.*

So, other things being equal, an electron in a state of high angular momentum tries to keep farther from the nucleus than an electron in a state of low angular momentum. This shows that the higher the value of angular momentum of an electron, there is lesser likelihood of its being found close to the nucleus. Thus, for a given values of *n*, the size of the orbit increases as *l* increases.

Space Quantisation of Electronic Orbits

When the electric current flows in a wire bent into the form of a ring is a magnetic dipole is produced perpendicular to the plane of the ring. The magnitude of the dipole is given by Eq. (5.104).

$$|\mu| = \text{current} \times \text{area of the ring} \qquad \text{...(5.109)}$$

If, current is expressed in electrostatic units and the magnetic dipole in electromagnetic units, we have

$$|\mu| = \frac{\pi I a^2}{c} \qquad \text{...(5.110)}$$

where I is the current strength, a the radius of the ring, and c the velocity of light in cm/sec. An electron with charge e and moving with a linear velocity v, in a circular orbit of radius a, can be considered equivalent to a circular current I, given by

$$I = \frac{ev}{2\pi a} \qquad \text{...(5.111)}$$

Thus

$$|\mu| = \frac{ev}{2\pi a} \times \frac{\pi a^2}{c} = \frac{eva}{2c}$$

$$= \frac{evm_e a}{2m_e c} = \frac{e}{2m_e c}|L|$$

or

$$\mu = \frac{e}{2m_e c} L \qquad \text{...(5.112)}$$

The axis of the magnetic dipole is antiparallel to the angular momentum vector due to the pressure of negative charge on the electron.

So, the magnitude of the total angular momentum is given by $\sqrt{l(l+1)}\, h/2\pi$. Thus

$$\mu = \sqrt{l(l+1)}\, \beta_e \qquad \text{...(5.113)}$$

where

$$\beta_e = \frac{eh}{4\pi m_e c} \qquad ...(5.114)$$

The magnetic unit of magnetic moment is called Bohr magneton.

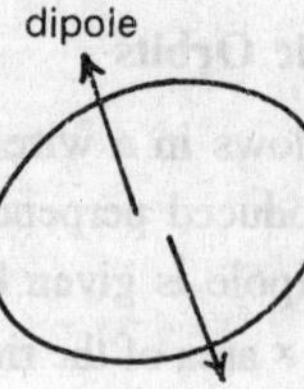

Fig. 5.7. *Relative orientation of the axes of magnetic dipole and rotation.*

As the operator for the component of angular moments along any chosen direction, say $\hat{L}_z$, has the eigenvalues $mh/2\pi$, where m can take $(2l + 1)$ different internal values in the range $-l \leq m \leq +l$, we can write

$$\cos\theta = \frac{m}{\sqrt{l(l+1)}} \qquad ...(5.115)$$

where θ is the angle between the axis of rotation and the chosen direction Z. The equation shows that $|m|$ cannot be more than $|l|$, otherwise $\cos\theta$ becomes greater than ± 1. This equation also shows that when $m \pm l$, $\cos\theta$ does not equal ± 1 and so, the axis of rotation never coincides, with the Z-axis. The component of the magnetic moment in any chosen direction is given by

$$|\mu| \cos\theta + m\,\beta_0 \qquad ....(5.116)$$

There are $(2l + 1)$ different components of magnetic moment, we expect that the magnetic moment vectors should make all possible angles with any chosen direction. Quantum mechanically, the magnetic moment vectors and hence the electronic orbits can make only certain discrete orientations in space. This is called the space quantisation of electronic orbits Fig. (5.8).

These magnetic moment vectors having certain discrete orientations in space refer to the same energy in absence of magnetic field. If we put a magnetic dipole μ in a magnetic field H, the energy of the dipole change in the amount.

$$\Delta E = -\mu H = -|\mu|\,|H| \cos\theta \quad \text{...(5.117)}$$

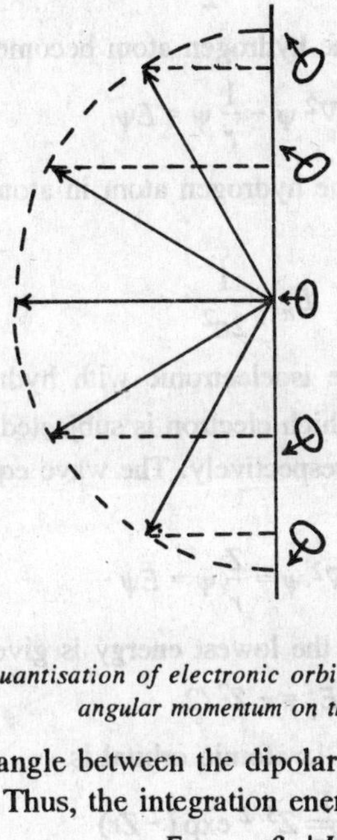

Fig. 5.8. *Space quantisation of electronic orbits and the various projections of angular momentum on the Z-axis.*

where θ is the angle between the dipolar axis and the direction of the magnetic field. Thus, the integration energy Δ E may be put as

$$E = -m\,\beta_e\,|H| \quad \text{...(5.118)}$$

Hence, the magnetic moment vectors having the different orientations with respect to the external magnetic field have different energies which are determined by the quantum number *m*. Due to such association with the magnetic field, the integer *m* is known the magnetic quantum number.

Energy of the Hydrogen Atom and the Isoelectronic Ions

The minimum energy which is needed to remove from the hydrogen atom in its ground state is known as the first ionization potential. It is equal to the negative of the first eigenvalue, *i.e.*

$$E_1 = -\frac{2\pi^2 m_e e^4}{h^2} = -13.6\text{ eV} \quad \text{...(5.119)}$$

In the system of atomic units, the radius of the first Bohr orbit of the hydrogen atom is considered to have unit length and $e^2 = 1$, so, Eq. (5.115) may be written as

$$E_1 = -\frac{1}{2} \qquad \text{....(5.120)}$$

and the wave equation for the hydrogen atom becomes

$$-\frac{1}{2}\nabla^2 \psi - \frac{1}{r}\psi = E\psi \qquad \text{...(5.121)}$$

The energy levels of the hydrogen atom in atomic units can be written

$$E_n = \frac{1}{2n^2} \qquad \text{...(5.122)}$$

Some atoms which are isoelectronic with hydrogen atom are, He^+, Li^{++}, Be^{+++}, etc., in which electron is subjected to the potential Z/r where Z is 2, 3, 4, etc., respectively. The wave equation for these system can be written as

$$-\frac{1}{2}\nabla^2 \psi = \frac{Z}{r}\psi = E\psi \qquad \text{...(5.123)}$$

where the first eigenvalue or the lowest energy is given as

$$E_1 = -Z^2/2 \qquad \text{...(5.124)}$$

and the wave function of the 1s atomic orbital is

$$\psi_{1s} = \frac{1}{\sqrt{\pi}} Z^{3/2} \exp(-Zr) \qquad \text{....(5.125)}$$

The most probable distance between the electron and the nucleus can be found for the hydrogen atom and is given by

$$\bar{r} = \frac{1}{Z} \qquad \text{...(5.126)}$$

This shows that as the atomic number increases the orbital contracts. Also, the first ionisation potential increases as Z increases. The observed first ionisation potentials of He^+, Li^{++}, and Be^{+++} ions, are:

$$He^+ = 54.16 \text{ eV}$$

$$Li^{++} = 121.8 \text{ eV}$$

$$Be^{+++} = 216.6 \text{ eV}$$

SOLVED PROBLEMS

Example 1. *Create the real functions from the functions defined as follows for m = ± 2.*

$$\Phi(\phi) = \frac{1}{\sqrt{2\pi}} e^{im\phi}$$

Solution :

For $m = 2_s$ we have

$$\Phi_{m-2} = \frac{1}{\sqrt{2\pi}} e^{2i\phi} \quad ...(i)$$

$$\Phi_{m-2} = \frac{1}{\sqrt{2\pi}} e^{-2i\phi} \quad ...(ii)$$

Taking the linear combinations of Eqs. (*i*). and (*ii*), we get

$$\Phi_{+} = \frac{1}{\sqrt{2}} \frac{2}{\sqrt{2\pi}} \left[e^{2i\phi}\, e^{-2i\phi} \right]$$

$$= \frac{2}{\sqrt{2}\,\sqrt{2}} \cos 2\phi$$

$$= \sqrt{\frac{1}{\pi}} \cos 2\phi$$

$$\Phi_{-} = \frac{1}{\sqrt{2}} \frac{1}{\sqrt{2\pi}} \left[\frac{e^{2i\phi} - e^{-2i\phi}}{i} \right]$$

$$= \frac{2}{\sqrt{2}\,\sqrt{2\pi}} \sin 2\phi$$

$$= \sqrt{\frac{1}{n}} \sin 2\phi$$

The additional $\frac{1}{\sqrt{2}}$ is an extra normalisation factor.

Example 2. *Calculate the average distance of 1s electron from the nucleus when the wave function for 1s orbital of hydrogen atom is*

$$\psi_{100} = \frac{1}{\sqrt{\pi a^{03}}} e^{-r/a_0}$$

Solution :

The average distance of electron in 1*s* orbital can be taken as follows :

$$\langle r \rangle = \int_{\text{all space}} \psi^* \hat{r} \psi d_\tau$$

$$= \int_0^\infty \int_0^\pi \int_0^{2\pi} \left\{\frac{1}{\pi a_0^3}\right\}^{\frac{1}{2}} e^{-r/a_0}$$

$$r \left(\frac{1}{\pi a_0^3}\right)^{\frac{1}{2}} e^{-r/a_0} r^2 \sin\theta \, dr \, d\theta \, d\phi$$

$$= \frac{1}{\pi a_0^3} \int_0^\infty r^3 e^{-2r/a_0} \int_0^\pi \sin\theta \, d\theta \int_0^{2\pi} d\phi$$

$$= \frac{1}{\pi a_0{}^3} \left[\frac{3\,!}{(2/a_0)^{3+1}}\right] [-\cos\theta]_0^\pi \, [\phi]_0^{2\pi}$$

$$= \frac{3}{3} a_0 \quad \textbf{Ans.}$$

Example 3. *The normalised wave function for 1s electron may be put as follows:*

$$\psi_1 s = \left(\frac{1}{\pi a_0^3}\right)^{1/2} e^{-r/a_0}$$

Find out the probability of the electron in a sphere of radius $r = a_0$. *(K. K. 1976)*

Solution.

The probability can be put as under :

$$P = \int \psi^* \psi \; de$$

$$= \frac{1}{\pi a_0^3} \int_0^{a_0} e^{-2r/a_0} r^2 dr \int_0^\pi \sin\theta \, d\theta \int_0^{2\pi} d\phi$$

or

$$P = \frac{4_0}{a_0^3} \int_0^{a^3} e^{-2r/a_0} r^2 dr$$

Integrating this equation by parts, we obtain

$$P = \frac{4}{a_0^3} \left\{ \left[\frac{a_0}{2} r^2 e^{-2r/a_0} \right]_0^{a0} + \right.$$

$$\left[-\frac{a_0}{2}re^{-r/a}\right]_0^{a_0} + \left[-\frac{a_0^3}{4}e^{-r/a_0}\right]_0^{a_0}\Bigg\}$$

$$= 5e^{-} + 1 = 0.323 \qquad \textbf{Ans.}$$

Example 4. *The wave-function of the hydrogen atom for 1s state is*

$$\psi(1s) = \frac{1}{\sqrt{\pi}}\left(\frac{1}{a_0}\right)^{3/2} e^{-r/a_0}$$

where $a_0 = \frac{\hbar^2}{me^2}$ *is the Bohr's radius. Calculate the expectation value of the potential energy of electron in 1s state (only the algebraic expression is required, not the numerical answer).*

Solution:

The expectation value of function f in the state ψ can be put as follows:

$$\langle f \rangle = \frac{\int \psi^* f \psi \, d_\tau}{\int \psi^* \psi \, d_\tau}$$

The potential energy of the electron in the hydrogen atom can be put as follows:

$$V(r) = -\frac{e^2}{r}.$$

So the expectation value of potential energy of the electron in the hydrogen atom in 1s state may be put as follows:

$$\langle V \rangle = \frac{\int \psi(1s)\left(-\frac{e^2}{r}\right)\psi(1s)\, d_\tau}{\int \psi^*(1s)\,\psi(1s)\, d_\tau}$$

Here $\psi^*(1s) = \psi(1s)$

$$= \frac{1}{\sqrt{\pi}}\left(\frac{1}{a_0}\right)^{3/2} e^{-r/a_0}$$

and $\qquad d_\tau = dr\, r \sin\theta\, d\phi = r^2\, dr \sin\theta\, d\theta\, d\phi$

$$\therefore \quad \langle V \rangle = \frac{\iiint \left(-\frac{e^2}{r}\right) \frac{1}{\pi a_0^3} e^{-r/a_0} r^2 \, dr \sin\theta \, d\theta \, d\phi}{\iiint \frac{1}{\pi a_0{}^3} e^{-2r/a_0} r^2 \, dr \sin\theta \, d\theta \, d\phi}$$

$$= -\frac{e^2 \int_{-\infty}^{+\infty} e^{-r/a_0} r \, dr \int_0^{\pi} \sin\theta \, d\theta \int_0^{2\pi} d\phi}{\int_{-\infty}^{+\infty} e^{-2/a_0} r^2 \, dr \int_0^{\pi} \sin\theta \, d\theta \int_0^{2\pi} d\phi}$$

$$= -\frac{e^2 \int_{-\infty}^{+\infty} e^{-2r/a_0} r^2 \, dr}{\int_{-\infty}^{+\infty} e^{-2t/a_0} r^2 \, dr}$$

But $$\int_0^{\infty} e^{-\alpha x} x^2 \, dx = \frac{n!}{a^{n+1}}$$

$$\langle V \rangle = -\frac{e^2 \cdot 2 \cdot \frac{1!}{(2/a_0)^2}}{2 \cdot \frac{2!}{(2/a_0)^3}} = -\frac{e}{a_0}$$

$$= -\frac{e^2}{\left(\frac{me^2}{\hbar^2}\right)} = -\frac{me^4}{rh^2}$$

6

Quantum Chemistry and Valency

Introduction

The treatment of molecules according to the principles of wave mechanics is more difficult and therefore more approximate as compared to that of many electron atoms. The two main types of approximations are: (*i*) the valence bond approximations, and (*ii*) the molecular orbital approximations.

In variation method we have a means of approximating the energy of a system but for that we have to choose a trial wavefunction which is not always an easy task. For calculation of molecular energy levels we have the above mentioned two basic approaches. These will be referred to as *valence bond theory* and *molecular orbital theory.*

Valence bond theory was developed much earlier than molecular orbital theory but these days molecular orbital theory has become more popular because of its greater conceptual and mathematical simplicity. However for some aspects of molecular structure valence bond theory offers a simpler pictorial representation. This approach, therefore, is used more frequently for qualitative considerations.

The Valence Bond Theory

Lewis concept of an electron-pair bond in quantum mechanical expression forms the basis of the valence bond theory. Just like the molecular orbital theory, the valence bond theory is also an approximation method. In its basic approach to the structure of a molecule this theory is closely related to our conventional ideas of localized chemical bonds. In this approach, the atoms are considered to retain their individuality, and the bond arises from the interaction of the valence electrons when the atoms come closer. This view is more readily adapted to qualitative arguments than that presented by the molecular orbital theory, and for this reason the valence bond theory is more familar. In spite of this desirable feature, the valence bond method many a times leads to more complicated methematics than does the molecular orbital theory. Moreover, the determination of a trial function for a complex molecule is not as simple as it is in the moleculer orbital theory.

The Hydrogen Molecule

The ideas of the valence bond method can be appreciated to some extent by an analysis of the quantum mechanical treatment of the hydrogen molecule. Heitler and London in 1927 treated this problem successfully and their approach was basically that of the valence bond method. The trial wavefunction was somewhat different in the simple valence bond theory than it is in the simple molecular orbital theory. According to the molecular orbital theory, the wavefunction for electron (1) is made up of a linear combination of the hydrogen 1*s* orbitals just as it in the hydrogen molecule ion. Thus

$$\psi_1 = \psi_a(1) + \psi_b(1)$$

But, in the hydrogen molecule there are two electrons, and a one-electron molecular orbital wavefunction for the second electron will be given by

$$\psi_2 = \psi_a(2) + \psi_c(2)$$

This gives the configurational wave function

$$\psi_{MO} = \psi_1\psi_2 = \{\psi_a 1 + \psi_b 1\}\{\psi_h 2 + \psi_b 2\} \qquad ...(6.1)$$

In this case it can be seen that the configurational wave function can be expressed as a product of one-electron molecular orbital wavefunctions.

A basically different construction is used in the valence bond method. In it is suggested that if two unconnected systems can be separately described by wavefunctions ψ_a and ψ_b, respectively, then the wavefunction for the combined system will be given by

$$\psi_1 = \psi_a\psi_b \qquad ...(6.2)$$

Thus for the hydrogen molecule

$$\psi_1 = \psi_1(1)\psi_{1b}(2) \qquad ...(6.3)$$

This wave function given by Eq. (6.3) places electron (1) on atom *a* and electron (2) on atom *b*.

Although the potential energy diagram resulting from this wave function has a minimum, thus indicating a stable molecule the agreement with exepriment is not good. However, a much better trial function can be obtained by introducing a fundamental principle of quantum mechanics. In the wave function ψ_1 we have arbitrarily placed electron (1) on atom *a* and electron (2) on atom *b*. There is no basis for such a choice, and it is equally valid to reverse this arrangement, giving

$$\psi_{11} = \psi_{1sa}(2)\psi_{1sb}(1) \quad ...(6.4)$$

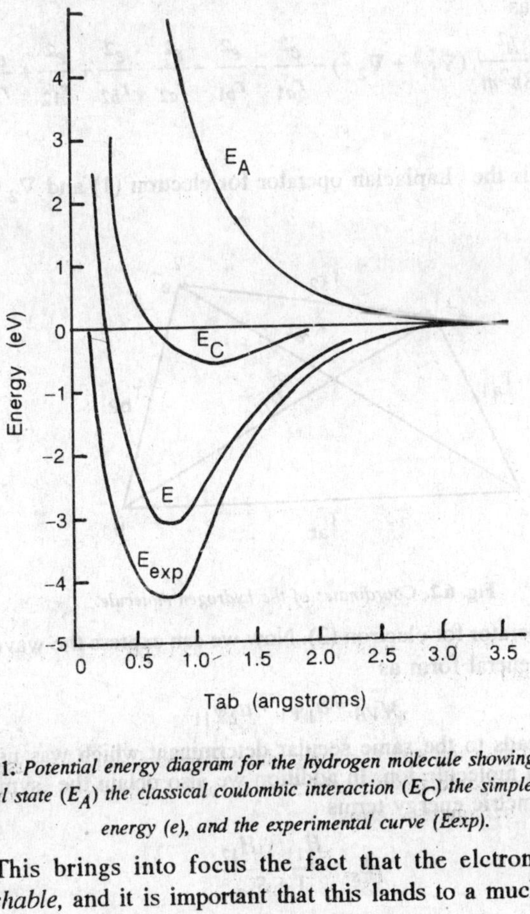

Fig. 6.1. *Potential energy diagram for the hydrogen molecule showing the antisymmetrical state (E_A) the classical coulombic interaction (E_C) the simple valence bond energy (e), and the experimental curve (Eexp).*

This brings into focus the fact that the elctrons are *indis tinguishable*, and it is important that this lands to a much better trial function. This is clear from Fig. 6.1, where the potential energy diagram for the hydrogen molecule is plotted for different trial functions. This new effect can be introduced into the total wavefunction if we take the usual linear combination according to the principle of supperposition, Thus we get

$$\psi_{VB} = \psi_{1sa}(1)\,\psi_{1sb}(2) + \psi_{1sa}(2)\psi_{1sb}(1) \qquad ...(6.5)$$

Symmetric and Antisymmetric Energy States

Now that we have a trial wavefunction, the problem is essentially reduces to that of the hydrogen molecule ion. Making use if the symbolism of Fig. 9.1, the Hamiltonian operator for the hydrogen moleucle is obtained as

$$H = \left[-\frac{h^2}{8\pi^2 m}(\nabla_1{}^2 + \nabla_2{}^2) - \frac{e^2}{r_{a1}} - \frac{e^2}{r_{b1}} - \frac{e^2}{r_{a2}} - \frac{e^2}{r_{b2}} + \frac{e^2}{r_{12}} + \frac{e^2}{r_{ab}}\right] \qquad ...(6.6)$$

where $\nabla_1{}^2$ is the Laplacian operator for electron (1) and $\nabla_2{}^2$ is the

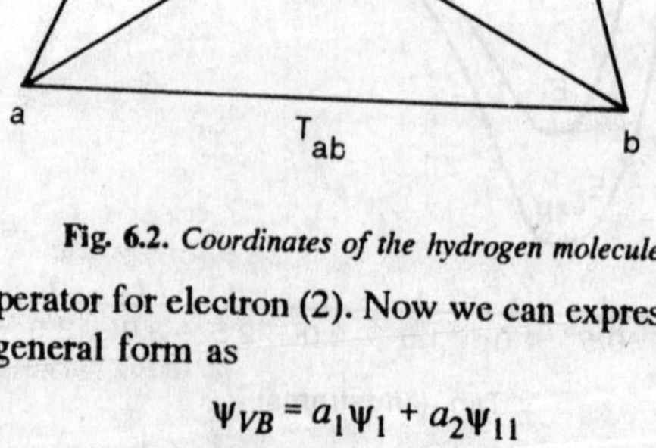

Fig. 6.2. *Coordinates of the hydrogen molecule.*

Laplacian operator for electron (2). Now we can express the wave function in the general form as

$$\psi_{VB} = a_1\psi_1 + a_2\psi_{11} \qquad ...(6.7)$$

This leads to the same secular determinant which was used for the hydrogen molecule ion. In addition we also obtain the symmetric and antisymmetric energy terms

$$E_s = \frac{H_{11} + H_{12}}{1 + S_{12}} \qquad ...(6.8)$$

and
$$E_A = \frac{H_{11} + H_{12}}{1 - S_{12}} \qquad \text{...(6.9)}$$

But the resultant integrals, will be more complex than those encountered in the hydrogen molecule ion.

The integral H_{11} is now defined as

$$H_{11} = \iint \psi_1 H\psi_1 \, dr_1 dr_2$$

The double integral arises due to a consideration of the co-ordinates of both electron (1) and electron (2). On introducing the Hamiltonian operator the integral becomes

$$H_{11} = \iint \psi_a(1)(2)\psi_b \left\{ -\frac{h^2}{8\pi^2 m}(\nabla_1{}^2 + \nabla_1{}^2) - \frac{e^2}{r_{a1}} - \frac{e^2}{r_{b1}} - \frac{e^2}{r_{a2}} - \frac{e^2}{r_{b2}} + \frac{e^2}{r_{12}} + \frac{e^2}{r_{ab}} \right\} \psi_a(1)\,\psi_b(2)\, dr_1 dr_2$$

Now,
$$\left(-\frac{h^2}{8\pi m}\nabla_1{}^2 - \frac{e^2}{r_{a1}} \right)\psi_a(1) = E_0\psi_a(1) \qquad \text{...(6.10)}$$

and
$$\left(-\frac{h^2}{8\pi^2 m}\nabla_2{}^2 - \frac{e^2}{r_{b2}} \right)\psi_b(2) = F_0\psi_b(2) \qquad \text{...(6.11)}$$

where E_0 is the ground-state energy of the hydrogen atom. Hence, the integral H_{11} can be expressed more simply as

$$H_{11} = \iint \psi_a(1)\,\psi_b(2) \left(2E_0 = \frac{e^2}{r_{ab}} + \frac{e^2}{r_{12}} - \frac{e^2}{r_{a2}} - \frac{e^2}{r_{b1}} \right) \psi_a(1)\psi_b(2)\, dr_1 dr_2 \qquad \text{...(6.12)}$$

Or, since the 1*s* wavefunctions are considered to be normalized

$$H_{11} = 2E_0 + \frac{e^2}{r_{ab}} + J_1 - 2J_2 \qquad \text{...(6.13)}$$

Here,
$$J_1 = e^2 \iint \frac{1}{r_{12}} [\psi_a(1)\,\psi_b(2)^2\, dr_1 dr_2] \qquad \text{...(6.14)}$$

and because of the equivalence of the two electrons, we have

$$J_2 = e_2 \iint \frac{1}{r_{a2}} [\psi_a\, 1\, \psi_b\, 2]^2\, dr_1 dr_2 \qquad \text{...(6.15)}$$

$$= e^2 \iint \frac{1}{c_{b_1}} [\psi_a(1)\psi_b 2]^2\, dr_1 dr_2 \qquad \text{...(6.16)}$$

The solution of these intergrals is even more complicated than that for the corresponding integrals of the hydrogen molecule ion, and therefore they will not be treated further. However, it will again be useful to consider the basic form of the energy expression.

By the same general argument as was used for H_{11}, we can also show that

$$H_{12} = E_0 S_{12} + \frac{e^2}{r_{ab}} + K_1 - 2K_2 \qquad \text{....(6.17)}$$

where $$K_1 = e^2 \iint \frac{1}{r_{12}} [\psi_a(1)\psi_b(2)\, \psi_a(2)\psi_b(1)\, dr_1 d\tau_2 \qquad \text{...(6.18)}$$

and, again due to the equivalence of the electrons,

$$K_1 = e^2 \iint \frac{1}{r_{a1}} [\psi_a(1)\psi_b(2)\, \psi_a(2)\psi_b(1)\, dr_1 d\tau_2 \qquad \text{...(6.19)}$$

$$= e^2 \iint \frac{1}{r_{b2}} [\psi_a(1)\psi_b(2)\psi_a(2)\, \psi_b(1)\, dr_1 d\tau_2 \qquad \text{...(6.20)}$$

Substituting these results into the expressions for the energy states of the hydrogen molecule and rearranging we can obtain

$$E_s - 2E_0 = \frac{e^2}{r_{ab}} + \frac{J_1 - 2J_2 + K_1 - 2K_2}{1 + S_{12}} \qquad \text{...(6.21)}$$

$$E_A - 2E_0 = \frac{e^2}{r_{ab}} + \frac{J_1 - 2J_2 - K_1 + 2K_2}{1 - S_{12}} \qquad \text{...(6.22)}$$

that correspond to the symmetric and the antisymmetric wavefunctions, respectively.

The potential energy curves obtained from both the symmetric and the antisymmetric states are seen in Fig. 6.1. By setting the ground-state energies of the isolated hydrogen atoms equal to zero, that is, $E_0 = 0$, then the resultant potential energy curve exhibits the *interaction energy* between the two hydrogen atoms as they form the hydrogen molecule. The antisymmetric state fails to show a minimum and as expected is represents an unstable state. But the symmetric state exhibits a relatively deep minimum in the curve, thereby indicating a stable molecular species. It is important that the potential minimum using this trial function is considerably closer to the experimental potential minimum than that obtained by ignoring the indistinguishability of the electrons. However, still better agreement can be obtained by using still other wavefunctions.

The Classical Interaction Energy

The indistinguishability of the electrons in the hydrogen molecule can be compared to admitting the existence of the two structures

$$H_a(1) \quad _{\text{I}} \quad H_b(2) \quad \text{and} \quad H_a(2) \quad _{\text{II}} \quad H_b(1)$$

Due to the identity of any two isolated hydrogen atoms, the wavefunctions of the two structures are identical as also the energies of the two structures. Now, to appreciate the significance of the indistinguishability of the two electrons, it is essential to interpret the molecular energy in terms of either one or the other of these structures. For example if we consider structure I, then the corresponding wave function is

$$\psi_1 = \psi_a(1)\,\psi_b(2) \qquad \text{...(6.23)}$$

and the energy of the system is given by

$$E = \frac{\int \psi H \ \psi d\tau}{\int \psi d\tau}$$

$$= \frac{\int \psi_a(1)\,\psi_b(2)\ H\ \psi_a(1)\psi_b(2)\ d\tau}{\int \psi_a(1)\psi_b(2)\psi_a(1)\psi_b(2)\ d\tau}$$

or

$$E_c = \frac{H_{11}}{S_{11}} \qquad \text{...(6.24)}$$

Using normalized $1s$ wavefunctions, it can be simplified to get $S_{11} = 1$, and

$$E_c = H_{11} \qquad \text{...(6.25)}$$

we have already shown that

$$H_{11} = 2E_0 + \frac{e^2}{r_{ab}} + J_1 - 2J_a \qquad \text{...(6.26)}$$

So, we can say that

$$E_c = 2E_0 + \frac{c^2}{r_{ab}} + J_1 - 2J_2 \qquad \text{...(6.27)}$$

The interaction energy of the two atoms, can be found by equating the energy of the isolated atoms, E_0, to zero. The resultant potential energy diagram then gives the decrease in energy of the system in comparison to that of the isolated atoms when the two hydrogen atoms approach each other. If a potential minimum as a function of r_{ab} is seen then the theory would predict a stable molecule. In Fig. 6.1, the second line represents E_c, and it shows a slight minimum in the potential diagram. This indicates that the molecule is stable, but the calculated dissociation energy is only about one-tenth of the experimental value.

If we consider the various terms in the energy expression then we can conclude the E_c is essentially the classical Coulombic interaction between the two hydrogen atoms. The integral J_1 is given by

$$J_1 = e^2 \iint \frac{1}{r_{12}} \psi_a^2(1)\, \psi_b^2(2)\, d\tau_1\, d\tau_2$$

The term c^2/r_2 shows the elctrostatic repulsion between two unit charges and $\psi_a{}^2$ (1) and $\psi_b{}^2$ (2) represent the charge densities of electron (1) and electron (2) respectively. It can be concluded that J_1 represents the elctrostatic repulsion between the charge clouds of electron (1) and electron (2).

The integral J_2 is given by

$$J_2 = e^2 \iint \frac{1}{r_2}\ \psi_a(1)\psi_b(2)\psi_a(1)\psi_b(2)\, d\tau_1 d\tau_2 \quad \text{...(6.28)}$$

$$= c^2 \iint \frac{1}{r_{b1}}\ \psi_a(1)\psi_b(2)\psi_a(1)\psi_b(2)\, d\tau_1 d\tau_2 \text{...(6.29)}$$

and these reduce effectively to

$$J_2 = e^2 \int \frac{1}{r_{a2}}\ \psi_b^2(2)\, d\tau \quad \text{...(6.30)}$$

$$= c^2 \int \frac{1}{r_{i1}}\ \psi_a{}^2(1)\, d\tau_1$$

This represents the attraction between the electron charge around each individual atom and the opposite nucleus. For example, it is the attraction between the positively charged nucleus b and the negatively charged electron cloud nucleus a.

Finally, e^2/r_{ab} represents the nuclear repulsion between the two positively charged nuclei. Thus, we have the two electrostatic repulsion terms, e^2/r_{ab} and J_1, and we have the one electrostatic attraction term $2J_2$. E_c should then represent the classical electrostatic attraction between the two hydrogen atoms. This leads to a stable hydrogen molecule, but it is not nearly so stable according to this picture as it is experimentally known to be.

Resonance

To appreciate the significance of the indistinguishability of the electrons we have to recognize that the two structures I and II are possible and a new wavefunction can result that leads to a dissociation energy which is only about 70% of the experimental value. Where as the simple electrostatic interaction gives a dissociation energy of only about 10% of the experimental value. It is thus understandable why the effect is considered a quantum mechanical principle. There is no classical analog of this effect, but in terms of our quantum mechanical model it has some meaning. In valence bond language, this effect is known as *resonance,* and it plays a critical role in the valence bond theory.

Mathematically we may conclude that electron (1) exists on atom *a* and electron (2) exists on atom *b* at any given instant, and at another instant the reverse arrangement exists. In this way the electrons could be considered to resonate between the two structures bus. This conclusion, is not the presently accepted conclusion. It is more realistic to say that the actual state of the system can be represented by neither structure I nor II at any instant of time, but the actual structure is some intermediate that has some of the character of each of the twc independent structures. Thus, we are constructing a wavefunction which describes the actual structure by introducing the various common characteristics expressed by each individual structure. Carrying out this construction leads to a much better wavefunction than that of either structure by itself and correspondingly, a significant increase in stability of the system with respect to any single structure. This difference in energy is known as the *resonance energy.*

It is significant to recognize that the concept of resonance is a *fictitious* concept and has its existence by virtue of the mode of construction of the trial wavefunction. The reality of this is thus a product of the quantum mechanical model used to describe the system. While making use of the valence bond theory, resonance has a reality, because it is basic to the language and concepts of this theory, but in terms of another equally good model, it may have no meaning and, hence no existence. This argument also applies to any scientific concept, but due to the success and use of some concepts they are generally given more credence than they actually deserve. Resonance belongs to this category. Thus, in terms of the valence bond theory, resonance will have an existence, and we will feel justified in speaking as if it were a fact. But in terms of other models, we may tend to its existence

Contribution of ionic terms

On the basis of the valence bond theory, the trial function can be improved by recognizing that two additional structures can also contribute to the total wavefunction of the hydrogen molecule. These are the ionic structures III and IV

$$\underset{\text{III}}{:H_b^- \quad H_a^+} \quad \text{and} \quad \underset{\text{IV}}{H_b^+ \quad :H_a^-}$$

In these both of the electrons are placed on either atom *a* or atom *b*. These structures therefore cannot be of equal weight with the homopolar structures I and II. Hence the total wavefunction can be represents as

$$\begin{aligned}\psi_{VB} &= a\psi_{I} + a\psi_{II} + b\psi_{III} + \psi b_{IV}\\ &= a\psi_a(2) + a\psi_a(2)\psi_b(1)\\ &\quad + b\psi_a(1)\psi_a(2) + b\psi_b(2)\end{aligned} \qquad ...(6.31)$$

Here the two homopolar terms are of equal weight and the two ionic terms are also of equal weight. If the atoms were not the same, this would not be true. So, in general, it can be said that

$$\psi_{VB} = a\psi_{I} + b\psi_{II} + c\psi_{III} + d\psi_{IV} + \ldots \qquad ...(6.32)$$

The values of the parameters *a, b, c, d,* etc., can be obtained by minimising the energy of the system with respect to each of the parameters. Such a calculation for the hydrogen molecule means that the bond has about 17% ionic character. Though the simple molecular orbital wavefunction for the hydrogen molecule contains the ionic terms represented by structures III and IV, yet they are given the same weight as the homopolar terms. This can be seen by miltiplying out the function

$$\psi_{MO} = \{\psi_a(1) + \psi_b(1)\psi_a(2) + \psi_b(2)\} \qquad ...(6.33)$$

to give

$$\psi_{MC} = \psi_a(1)\psi_b(2) + \psi_a(2)\psi_h(1) + \psi_a(1)\psi_a(2) + \psi_b(1)\,\psi_b(2) \qquad ...(6.34)$$

Hence it appears that the simple molecular orbital theory places too much emphasis on the ionic terms. This can be corrected by arbitrarily introducing parameters to give the proper weight to the different terms.

The valence bond function can be still futher improved by introducing other parameters. These are generally attributed to different physical aspects of the molecular structure. For example, we can represent the 1*s* wavefunction as $\psi = (1/\pi a_0{}^3)\, e^{-ar/a0}$ rather than $\psi = (1/\pi a^{03})\, e^{-a/a0}$. The parameter, α, can then be used to minimize the energy, and also it can be justified in terms of a contraction of the electron orbit due to an attraction to two nuclear centres. Thus, no justification is really necessary. More the parameters that are used in the trial function, more adaptable it should be, if we ignore the effort needed to solve the resultant equations. Actually essentially exact agreement between theory and experiment was obtained by James and Coolidge using a wave function with 13 terms.

Wave Functions of the Hydrogen Atom

In the Bohr atom, it was postulated that the electron follows a specified circular path in its motion about the nucleus. This specified path was the simplest path which gives agreement between the theoretical model and experiment. Such a circular path led to a symmetrical atom and contributed almost nothing to an understanding of the geometry of molecules. But the more outstanding accomplishments of the quantum mechanical approach to atomic structure has been its determination of the general distribution pattern of the electron about the atom, and the relationship of this pattern to molecular structure.

It was postulated that the square of the wavefunction is a measure of the probability distribution of the electron. This wavefunction is made up of two parts, an angular portion which is represented by $Y_{lm}(\theta, \varphi)$ and a radical portion which is represented by $R_{n,l}(r)$. The radial portion of the wavefunction gives the distribution of the electron with respect to its distance from the nucleus whereas the angular portion gives the geometry of the various electron states.

The normalized solutions of the θ equation and also the radial equation are, quite complex but they reduce to relatively simple forms on introduction, of particular values of the parameters. For the θ equation, the allowed values of the parameter *m* are seen to be m=0, ±1, ±2, . . ., ±*l*. These give the normalized functions of θ(φ) shown in Table 6.1, in which both the complex and the real forms are given. Examples of the normalized (θ) functions and radial functions are given in Tables 6.3 and 6.2, respectively.

The normalized total wave function for the hydrogen atom can be obtained by using the relation:

$$\Psi_{(r\theta\varphi)} = R_{n,l}(r)\, Y_{lm}(\theta_{1\varphi})$$

TABLE 6.1
Normalized Functions of $\Phi_m(\varphi)$

$$\Phi_m(\varphi) = \frac{1}{\sqrt{2\pi}} e^{\pm ime}$$

$$\Phi_0(\varphi) = \frac{1}{\sqrt{2\pi}} \quad \text{or} \quad \Phi_0(\varphi) = \frac{1}{\sqrt{2\pi}}$$

$$\Phi_1(\varphi) = \frac{1}{\sqrt{2\pi}} e_{i\varphi} \quad \text{or} \quad \Phi_1(\varphi) = \frac{1}{\sqrt{\pi}} \cos\varphi$$

$$\Phi_{-1}(\varphi) = \frac{1}{\sqrt{2\pi}} e^{-i\varphi} \quad \text{or} \quad \Phi_{-1}(\varphi) = \frac{1}{\sqrt{\pi}} \sin\varphi$$

TABLE 6.2
Normalized Functions of $\Phi_{l \cdot m}(\theta)$

$$\Phi_{im}(\theta) = \sqrt{\frac{(2i+i)l - |m|)l}{(l+|m|)\,!}} \; P_i\,|\,m\,|\,(\cos b)$$

$$l = 0: \quad \theta_{0,0}(\theta) = \frac{\sqrt{2}}{2}$$

$$l = 1: \quad \theta_{1,0}(\theta) = \frac{\sqrt{6}}{2} \cos\theta$$

$$\theta_{1}, \pm_{1} = \frac{\sqrt{3}}{2} \sin\theta$$

TABLE 6.3
Normalized Functions of $R_{n,l}(r)$

$$R_{n,l}(r) = \sqrt{\left(\frac{2Z}{na_0}\right)^2 \frac{(n-1-1)\,!}{2n\,[(n+l)\,!]^3}} \; e^{-\rho/2} \rho^l L_{n+1}^{2l+1}(\rho)$$

$n = 1$, K shell:

$$l = 0: \quad R_{1,0}(r) = (Z = a_0)^{3/2}\, 2e^{-\rho/2}$$

$n = 2$, L shell:

$$l = 0: \quad R_{2,0}(r) = \frac{(Z/a_0)^{3/2}}{2\sqrt{2}} (2-\rho)\, e^{-\rho/2}$$

$$l = 1: \quad R_{2,1}(r) = \frac{(Z/a_0)^{3/2}}{2\sqrt{6}} \rho e^{-\rho/2}$$

As can be seen, the particular choice of functions of $R_{n},\ _{1}(r)$, $\Phi_{l,m}(\theta)$, and $\Phi_{m}(\varphi)$ are not arbitrary. They are limited by the permitted values of the quantum numbers. By using all possible arrangements of these functions within the limitations of the quantum numbers, we obtain the normalized total wavefunctions as given in Table 6.4 where

$$\sigma = \frac{n}{2}\rho.$$

TABLE 6.4

Normalized Hydrogen-like Wavefunctions

K Shell:

$n = 1, l = 0, m = 0$: $\psi_{1s} = \frac{1}{\sqrt{\pi}}(Z/a_0)^{3/2} e^{-\sigma}$

L Shell:

$n = 2, l = 0, m = 0$: $\psi_{2s} = \frac{1}{4\sqrt{2\pi}}(Z/a_0)^{3/2}(2-\sigma)e^{-\sigma/2}$

$n = 2, l = 1, m = 0$: $\psi_{2px} = \frac{1}{4\sqrt{2\pi}}(Z/a_0)^{3/2}\sigma e^{-\sigma/2}\cos\theta$

$n = 2, l = 1, m = \pm 1$: $\begin{cases} \psi_{2px} = \frac{1}{4\sqrt{2\pi}}(Z/a_0\ ^{3/2}\sigma e^{-\sigma/2}\sin\theta\cos\varphi \\ \psi_{2py} = \frac{1}{4\sqrt{2\pi}}(Z/a_0)^{3/2}\sigma e^{-\sigma/2}\sin\theta\cos\varphi \end{cases}$

M Shell:

$n = 3, l = 0, m = 2$: $\psi_{2s} = \frac{2}{81\sqrt{3\pi}}(Z/a_0)^{3/2}(27 - 81\sigma + 2\sigma^2)\sigma e^{-\sigma/3}$

Radial Wavefunction

Let us normalize the ψ_{100} wavefunction.

$$\int_{-\infty}^{\infty} \psi_{100}^2\, d_\tau = 1,$$

where d_τ represents the volume element in which the electron is likely to be present. In polar co-ordinates

$$d_\tau = r^2 \sin\theta\, d\theta\, d\phi\, dr$$

we know that $\psi_{100} = Ae^{-\rho/2}$

$$\int_{\infty}^{\infty} \psi_{100}{}^2\, d_\tau = A^2 \int_0^{\infty}\int_0^{\pi}\int_0^{2\pi} e^{-Zr/a_0} r^2 \sin\theta\, d\theta\, d\phi\, dr$$

$$= 2.2\pi A^2 \int_0^\infty e^{-2Zr/a_0}\, r^2 dr$$

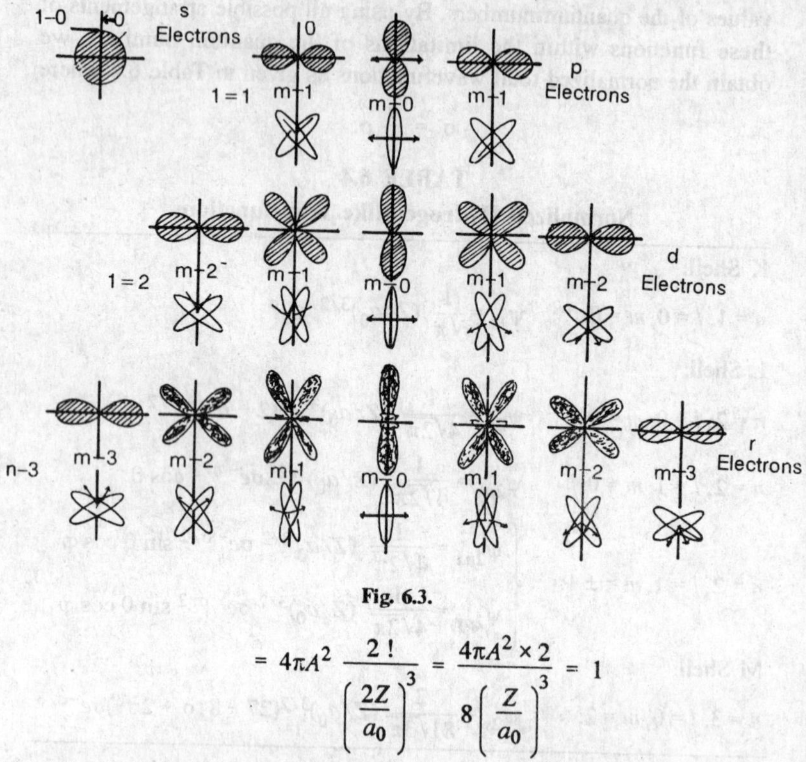

Fig. 6.3.

$$= 4\pi A^2 \frac{2\,!}{\left(\frac{2Z}{a_0}\right)^3} = \frac{4\pi A^2 \times 2}{8\left(\frac{Z}{a_0}\right)^3} = 1$$

or $$A^2 = \frac{1}{\pi}.\left(\frac{Z}{a_0}\right)^3$$

or $$A = \frac{1}{\sqrt{\pi}}\left(\frac{Z}{a_0}\right)^{3/2}$$

Therefore $\pi_{100} = \dfrac{1}{\sqrt{\pi}}\left(\dfrac{Z}{a_0}\right)^{3/2} e^{-Zr/a_0}$

For hydrogen atom, $Z = 1$, and so

$$\psi_{100} = \frac{1}{\sqrt{\pi a_0{}^3}}.\, e^{-r/a_0}$$

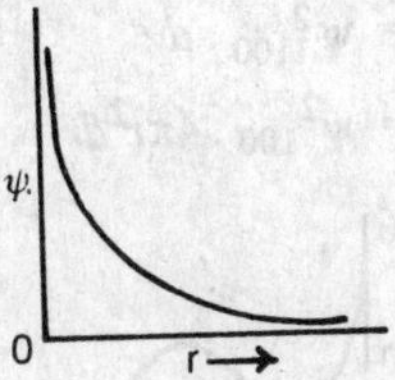

Fig. 6.4.

Now ψ^2 or $\psi\psi^*$ represents the probability distribution function for the electron relative to the nucleus. This is given as

$$\psi^2 = \frac{1}{\pi a_0^3} e^{-2r/a_0}$$

This is a quantity independent of θ or φ; hence the normal hydrogen atom will be spherically symmetrical. Since the Bohr orbit was restricted to a single plane, so the Bohr atom lacks in this symmetry.

Curves given in Fig. 6.4 and 6.5 show ψ_{100} and ψ^2_{100} as a function or *r*, the distance from the nucleus. It is clear that the pobability is maximum in the nucleus ($r = 0$). For functions of the type $\rho e^{-\rho/2}$ *i.e.* re^{-r}, e^{-r} falls exponentially. In practice both these opposing tendencies will contribute and hence the actual wavefunction distribution will be a curve of the type shown in the figures. *The density distribution does not mean that the electron is smeared into a cloud. It is still a point charge but its position and momenta can not be fixed simultaneously* and so electronic transition can not occur in this case.

Radial Distribution Function

Instead of the probability distribution funciton ψ^2, we may now consider the radial distribution function by which we mean the probability of finding the electron within a spherical shell of thickness *dr* at a distance *r* from the nucleus.

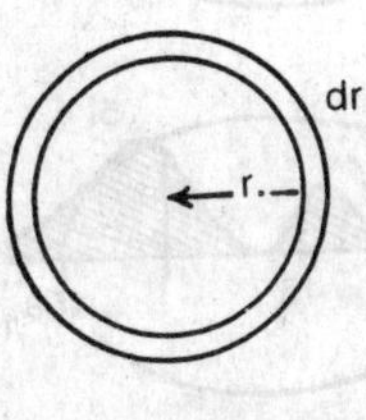

Fig. 6.6.

$$P(r)dr = \psi^2_{100} \cdot dv$$
$$= \psi^2_{100} \cdot 4\pi r^2 dr$$

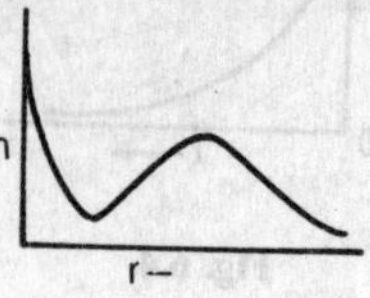

Fig. 6.7.

$$= \frac{1}{\pi} \cdot \left(\frac{Z}{a_0}\right)^3 e^{-2Zr/a_0} \cdot 4\pi r^2 \, dr$$

$$= 4\left(\frac{Z}{a_0}\right)^3 e^{-2Zr/a_0} \cdot v^2 \, dr$$

The functions are graphically represented as in Fig 6.6 and 6.7.

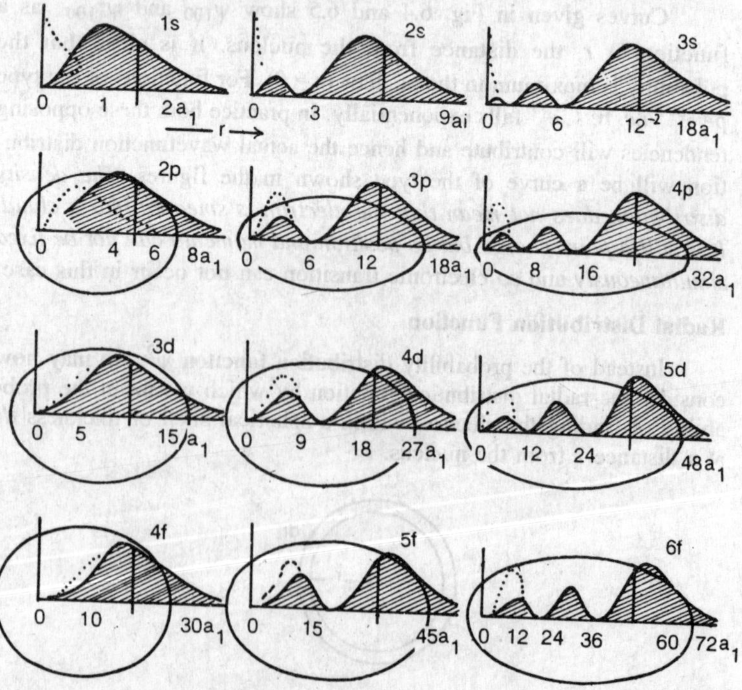

Fig. 6.8.

These correspond to a more diffuse distribution of electric charge and the maxima in these distribution curves correspond closely to the radii of Bohr orbits and for hydrogen the electron density attains a maximum at a value of r which is just the same as calculated by Bohr to be the first orbit or radius 0.529 Å. This can be established thus:

$$\frac{dP(r)}{dr} = 0 \text{ at maximum}$$

But $$P_r = Ar^2e^{-2Zr/a_0} \qquad \text{by (50.48)}$$

$$\frac{dP_r}{dr} = 2r\,Ae^{-2Zr/a_0} - \frac{2Z}{a_0}r^2A\,.\,e^{-2Zr/a_0} = 0$$

or $$1 - \frac{Zr}{a_0} = 0 \quad \text{or} \quad r = \frac{a_0}{Z} \quad \text{or} \quad r = 0.529 \text{ Å}$$

since $$Z = 1.$$

This distribution function is not similar to that for a_0 in circular Bohr orbit which would be everywhere zero except at $r = \psi_0$.

Radial Distribution Curves

In its ground state 'the wavefunction for the hydrogen atom is given by

$$\Psi_{1s} = \frac{1}{\sqrt{\pi a_0{}^2}}\,e^{-r/a_0}$$

The square of this function gives the probability distribution for the electron with respect to the nucleus. The fact that a probability distribution exists at all is in conflict with the ideas of the Bohr theory, where the electron was confined to a prescribed orbit at a given distance from the nucleus. Quantum mechanically, however, we see from the plot of the dependence of the radial wavefunction on the distance, shown in Fig. 6.9, that there is 2 finite probability for the electron to be even at very great distances from the nucleus. This probability, of course, is exceedingly small, but it does exist.

The elctron distribution can be expressed in terms of the radial distribution function which is a measure of the probability of finding the electron in a spherical shell between the distances r and $r + dr$ from the nucleus. The volume lying between two concentric spheres at a distance of r to $r + dr$ from the origin can be obtained by mul-

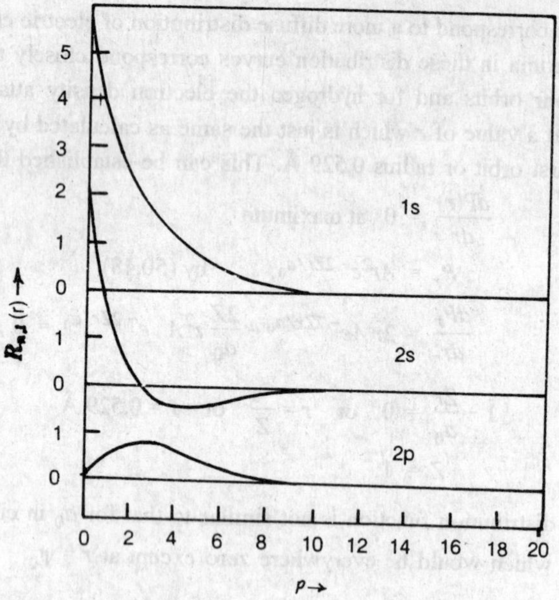

Fig. 6.9. Radial wave functions for the 1s, 2s, and 2p states of the Hydrogen Atom.

tiplying the surface area of a sphere, $4\pi r^2$, by the distance between it and the next concentric sphere, dr. The probability that the electron is in this volume element should be given by

$$D_{(r)}dr = 4\pi r^2 \psi\psi^* dr$$

$$D_{(r)}dr = \frac{4}{a_0^3} r^2 e^{-2r/\sigma_s} dr$$

Though the electron has a finite probability of existence at great distances from the nucleus yet its maximum probability is at a_0, exactly the distance calculated by Bohr for the electron orbit in the Bohr atom.

Antisymmetric Nature of Overall Wavefunctions

The overall wavefunction of an electron contains both orbital wavefunction and a term because of spin of the electron about its axis. The latter can be represented in terms of a spin wavefunction and the total can be represented as a product of orbital and spin wavefunctions. This is so as there is a very little interaction between the spin and the orbital angular momentum and hence between their wavefunctions so they can be treated separately. Now for the same n, l and m values, if the orbital wave function be ψ then the spin wavefunction will be either α or β depending on whether the spin is $\frac{1}{2}$ or $-\frac{1}{2}$. Hence the overall wavefunction will be $\psi\alpha$ or $\psi\beta$.

In the case of a molecule involving two nuclei and two electrons it has been found that two orbital wavefunctions arc obtained. One is ψ_σ (the symmetric wavefunctions) = $c_1 (\psi_1 - \psi_n)$ and other is ψ_A (the antisymmetric wavefunction) = $c_1 (\psi_1 - \psi_n)$ where c_1 stand for the normalising factor. Hence the rotal wavefunction will be the product of each of them and the total spin eigenfunction.

There can be four different values for the latter depending on whether the spin of electrons is $\frac{1}{2}$ or $-\frac{1}{2}$ or α and β.

ms_1	ms_2	$\Sigma\ ms^*$	*spin eigen*	*function*
$+\frac{1}{2}$	$+\frac{1}{2}$	1	$\alpha(1)$	$\alpha(2)$
$+\frac{1}{2}$	$+\frac{1}{2}$	0	$\alpha(1)$	$\beta(2)$
$-\frac{1}{2}$	$+\frac{1}{2}$	0	$\beta(1)$	$\alpha(2)$
$-\frac{1}{2}$	$-\frac{1}{2}$	-1	$\beta(1)$	$\beta(2)$

The second and the third terms differ only in exchange electrons and hence degenerate. By removing the degeneracy they can be expressed as the linear combination.

$$c_2[\alpha(1)\,\beta(2) + \beta(1)\,\alpha(2)]$$

and

$$c_2[\alpha(1)\,\beta(2) - \beta(1)\,\alpha(2)]$$

where c_2 = the normalising factor.

S_2 the total number of possible wavefunctions will be the product of ψ_s and ψ_A and each of this will form total spin wavefunction as

$$\psi_1 = \psi_S \cdot [\alpha(1)\alpha(2)]$$

$$\psi_2 = \psi_S \cdot [\alpha(1)\beta(2) + \beta(1)\alpha(2)]$$

$$\psi_3 = \psi_S \cdot [\alpha(1)\beta(2) - \beta(1)\alpha(2)]$$

$$\psi_4 = \psi_S \cdot [\beta(1)\beta(2)]$$

$$\psi_5 = \psi_A \cdot [\alpha(1)\alpha(2)]$$
$$\psi_6 = \psi_A \cdot [\alpha(1)\beta(2) + \beta(1)\alpha(2)]$$
$$\psi_7 = \psi_A \cdot [\alpha(1)\beta(2) - \beta(1)\alpha(2)]$$
$$\psi_8 = [\psi_A \cdot [\beta(1)\beta(2)].$$

The possiblity of the wavefunction cannot be given before hand. But from the spectrum of helium it has been found that the number of lines is such that for a system of many electrons, the total wavefunction must be antisymmetric for every pair of electrons. If the spin part is symmetric the orbital part would be antisymmetric or *vice versa.* But if both are symmetric or antisymmetric then the total will be symmetric. Hence only ψ_3, ψ_5, ψ_6, ψ_8 will give good wavefunctions.

In case the coordinates of the electrons are identical, then the antisymmetric orbital wavefunction will vanish and hence only ψ_3 will remain. Thus we find that for the same *n, l* and *m* value for the two-electron system, the spins are opposite. This is **Pauli's exclusion principle** which states that all four quantum nos. of two electrons in the same energy state cannot be equal *i.e.* if *n, l* and *m* are the same their spins must be opposite.

Corresponding to ψ_3, ψ_5, ψ_6 and ψ_8, ψ_3 corresponds to the symmetric wavefunction and it has been found only in Heitler-London treatment that the most stable molecule formation will occur when the orbital wavefunction is symmetric. So ψ_3 will give the wavefunction for the stable molecule of H_2. But ψ_5, ψ_6 and ψ_8 will give the unstable but the existent molecule. The energy level for ψ_3 state is undegenerate and so an energy level is to be obtained which is because small interaction of spin and orbital momenta. ψ_5, ψ_c and ψ_3 will give three entirely separated energy levels which differ from each other in that Σms is + 1, 0 or −1. Thus a triplet level is predicted.

Molecular Orbital Theory

The general concept of molecular orbitals and of the building up principle making use of molecular orbitals were developed by Hand and Mulliken in 1927 and by Lennard-Jones in 1929.

The LCAO Approximation

To get the wavefunction of ground state of a molecule the molecular orbitals are filled up with electrons starting from lowest energy and using various principles for filling up of the orbitals such as Pauli's exclusion, principle, Hund's rule etc.

The molecular orbitals are obtained by the linear combination of atomic orbitals which are associated with each atom in the molecule. Though each atom may contribute several atomic orbitals in the building up of atomic orbitals but for simiplicity we assume that each atom contributes only one orbital in n atomic molecule.

If $\phi_1, \phi_2, \phi_3 \ldots \phi_n$ be the atomic orbitals on atom 1, 2 . . . n respectively, then molecular orbitals are approximately written as

$$\psi = \sum_i^n C_i \phi_i \quad \ldots(6.35)$$

The electronic energy of this system is given by

$$E = \frac{\langle \sum_i^n anC_i\phi_i \mid He \mid \sum_i^n C_i \phi_i \rangle}{\langle \sum_i^n C_i \phi_i \mid \sum_i^n C_i \phi_i \rangle} = f(c_1, c_2 \ldots c_n) \quad \ldots(6.36)$$

As the function ψ in Eq. (6.35) has been chosen arbitrarily so the calculated energy of the system will always be greater than the true minimum energy (variation theorem).

The above energy Eq. (6.36) be minimised with respect to parameters $c_1, c_2, c_3 \ldots\ldots c_n$ The stationary values of E satisfy

$$\partial E(c_1, c_2, c_3, \ldots c_n)$$

$$= \frac{\partial E}{\partial c_1} \partial c_1 + \frac{\partial E}{\partial c_2} \partial c_2 + \ldots. + \frac{\partial E}{\partial c_n} \partial c_n = 0 \quad \ldots(6.37)$$

It gives the following solutions

$$\frac{\partial E}{\partial c_n} = 0, \qquad n = 1, 2, 3, \ldots n \quad \ldots(6.38)$$

These solutions will lead to approximations to the energy E_i and the corresponding molecuar orbital wavefunction ψ_1 for the stationary states.

From Eq. (6/36) we can write

$$E = \frac{\sum_i^n c_i^2 H_{ii} + 2 \sum_{i>j}^n c_2 cj H_{ij}}{\sum_i^n c_i^2 \Delta_{ii} + 2 \sum_{i>j}^n c_i cj \Delta_{ij}} \quad \ldots(6.39)$$

where

$$H_{ii} = \langle \phi_i \mid He \mid \phi_i \rangle$$
$$H_{ij} = \langle \phi_i \mid He \mid \phi_j \rangle$$
$$\Delta_{ii} = \langle \phi_i \mid \phi_i \rangle$$
$$\Delta_{ij} = \langle \phi_i \mid \phi_j \rangle$$

Rearranging Eq. (6.39), we get

$$\sum_i^n c_i^2 H_{ii} + 2 \sum_{i>j}^n c_i cj H_{ij} = E[\sum_i^n c_i^2 \Delta_{ii} + 2 \sum_{i>j}^n c_j \Delta_{ij}] \quad ...(6.40)$$

Differentiating Eq. (6.40) w.r.t. c_i, keeping other coefficients constant, we get

$$2 c_i H_{ii}^{+2} \sum_{J \neq i} cj H_{ij} = [2 c_i \Delta_{ii}^{+2} \sum_{J \neq i}^n c_i \Delta_{ij}]$$

$$+ \frac{\partial E}{jci} [\sum_i^n c_i^2 \Delta_{ii} + 2 \sum_{i>j} c_i cj \Delta_{ij} \quad ...(6.41)$$

For minimum energy, with respect to c_i, $\frac{\partial j}{\partial c_i} = 0$, so we have

$$2c_i (H_{ii} - \Delta_{ii} E) + 2 \sum_{j \neq i} ej (H_{ij} - \Delta_{ij}E) = 0$$

i.e.
$$\sum cj (H_{ij} - \Delta_{ij} E) = 0 \quad ...(6.42)$$

Thus if Eq (6.40) is minimised w.r.t. the parameters $c_1, c_2, c_3 \ldots$ etc. one by one, we obtain the following set of equations which are called *secular equations.*

$$C_1(H_{11} - \Delta_{11}E) + C_2(H_{12} - \Delta_{12}E) + C_3(H_{12} - \Delta_{13}E) + \ldots = 0$$
$$C_1(H_{21} - \Delta_{21}E) + C_2(H_{22} - \Delta_{22})E + C_3(H_{23} - \Delta_{23}E) + \ldots = 0$$

If the above set of n equations is to have non-trivial solutions the determinant of coefficients must vanish, that is

$$\begin{vmatrix} H_{11} - \Delta_{11}EH_{12} - \Delta_{12}EH_{13} - \Delta_{13}E \ldots H_{1n} - \Delta_{1n}E \\ H_{21} - \Delta_{21}H_{22} - \Delta_{22}E \ldots\ldots\ldots\ldots H_{2n} - \Delta_{2n}E \\ \downarrow \qquad \downarrow \\ H_{n1} - \Delta_{n1}EH_{n2} - \Delta_{n2} E \ldots\ldots\ldots\ldots H_{nn} - \Delta_{nn}E \end{vmatrix} = 0$$

L.H.S. of this determinant is known as *secular determinant.*

This approach forms the basis of LCAO treatment for calculating approximate energies and molecular orbitals in molecules.

Diatomic Molecules

In the construction of the one-electron molecular orbitals for the hydrogen molecule, we made use of linear combinations of the atomic ls orbitals of the isolated hydrogen atoms. In this case the isolated atoms are identical, and the ground state in each is the same but if a molecule is to be constructed from two different atoms then it is very unlikely that the same orbitals would be used from both atoms. For example if we consider the molecule HCl, it is clear that, although the 1*s* orbital of the hydrogen atom is used, the 1*s* orbital of the chlorine atom will not contribute to the formation of the bond. This emphasizes an important point ·vith regard to bond formation. Thus for two orbitals to form an effective molecular orbital, it is necessary that they be of comparable energies. In this case the 1*s* orbital of the chlorine atom is at a much lower energy than the 1*s* orbital of the hydrogen atom, and the two will, not combine to form a molecular orbital.

The extent of *overlap* between the combining orbitals is also importnat. Although the overlap is, by itself, not sufficient criterion for bond formation however it is important. Mathematically, the overlap can be expressed in terms of the overlap or orthogonality integral, $S_{ab} = \int \psi_A \psi_{13} \, d\tau$. When S_{ab} is large, the overlap of the orbitals ψ_A and ψ_B is large. The overlap criterion will be of extreme importance in case of directional bonding. If suffice to say that one of the factors that must be considered in the choice of atomic orbitals to be used in the construction of the molecular orbital.

On the same lines, it is important to consider the symmetry of the combining orbitals. We know that a *p* orbital has a positive and a negative lobe whereas a 1*s* orbital is everywhere positive. Now, as can

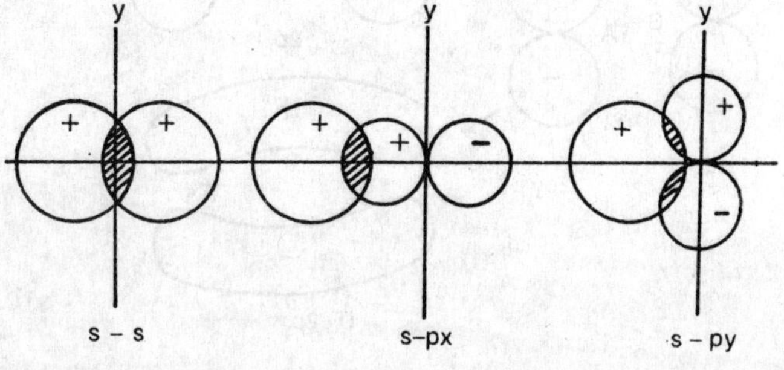

Fig. 6.10. *Overlap of s- and p-type orbitals with respect to axial symmetry.*

be seen in Fig. 6.10 two s orbitals overlap effectively as do an s and a p orbital. However, it is apparent that the symmetry of the p_y and p_z orbitals prevent an effective with the s orbital. We find that both the positive and the negative lobes of the p_y orbital overlap the s orbital and each exactly cancels the effect of the other. Similar is the situation with the p_z orbital. Just as no overlap at all causes $S_{ab} = 0$, the type of overlap observed with the p_y and p_z orbitals also given to $S_{ab} = 0$. Hence, it is clear that the relative energies, the extent of overlap, and the symmetry of the combining orbitals are considered in the construction of a molecular orbital wavefunction.

In the case of homonuclear diatomics, we can give a simple pictorial representation of the molecular orbitals as they are formed from

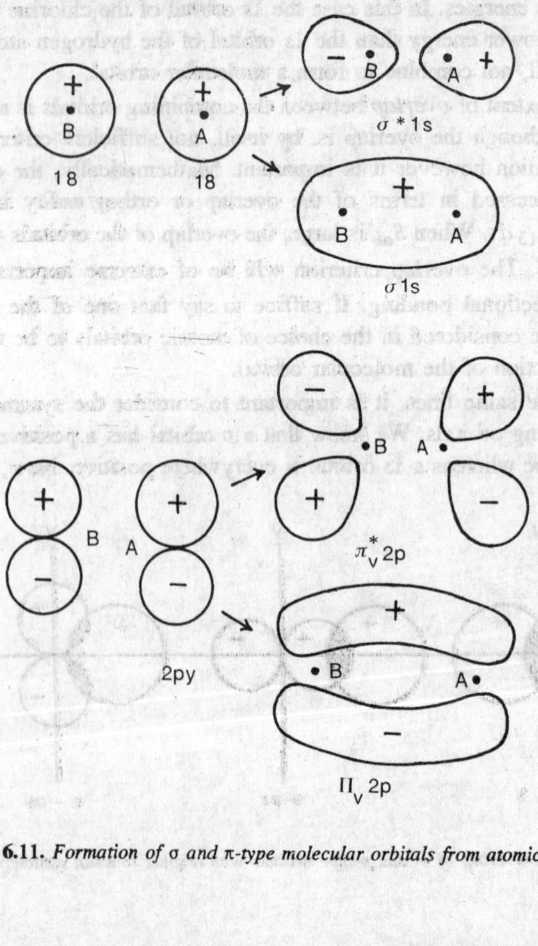

Fig. 6.11. *Formation of σ and π-type molecular orbitals from atomic orbitals.*

the individual atomic orbitals. For the *s* and *p* states, we can get two distinct types of molecular orbitals, as seen in Fig. 6.11. The atomic orbitals combine to form two types of molecular orbitals, the bonding and the antibonding molecular orbitals which correspond to the symmetric and the antisymmetric functions, respectively. The molecular orbitals formed from the 1s atomic orbitals have cylindrical symmetry about the line of centres between the two atoms, A and B. Such bonds are known as σ bonds and would also be expected from the combination of two p_x orbitals or an *s* with a p_x orbital. To the contrary, the p_y orbitals shown in Fig. 6.11 have a quite different symmetry with respect to the bond axis. The bonds are known as π bonds, and since they can be formed from either the p_y or the p_z atomic orbitals, they are designate as π_y and π_z, respectively.

In addition to the source of the orbitals, it is also essential to differentiate between the bonding and the antibonding orbitals. This is achieved by designating an antibonding σ orbital as σ* and an antibonding π orbital as π*. The same symbolism is used for δ orbitals, which originate from atomic *d* orbitals, and so on.

Like the filling of atomic orbitals, there is an order of filling of molecular orbitals. It has been found that for homonuclear diatomic molecules of the first-row elements the order is generally

$$\sigma_{1s} < \sigma^*_{1s} < \sigma_{2s} < \sigma^*_{2s} < \pi_{y2p} = \pi_{z_{2p}} < \sigma_{z_{2p}} < \pi^*_{z_{2p}}$$

$$= \pi^*_{z2p} < \sigma^*_{2p}$$

with the positions of the π_{2p} and σ_{2p} changing in the heavier members. Once this order is established, we can consider the formation of a homonuclear diatomic molecules between the first-row elements. This is done in much the same way as that used to build up the electron level in an atom. Consider the hydrogen molecule, it has two electron to fill up the molecular orbitals, and since each orbital can accommodate two electrons, we can represents the molecular formation as

$$\text{H}\,[1s^1] + \text{H}\,[1s^1] \rightarrow \text{H}_2\,[(\sigma_{1s})^2]$$

Thus, the hydrogen molecule has two electrons in the bonding σ_{1s} orbital that was constructed from the atomic 1*s* states of the isolated hydrogen atoms. As both electrons are in the bonding orbital so the resultant molecule should be stable. In case of hydrogen molecule the configurations is $H_2^+\,[(\sigma_{1s})]$, and the one electron in the bonding orbital should lead to a stable species.

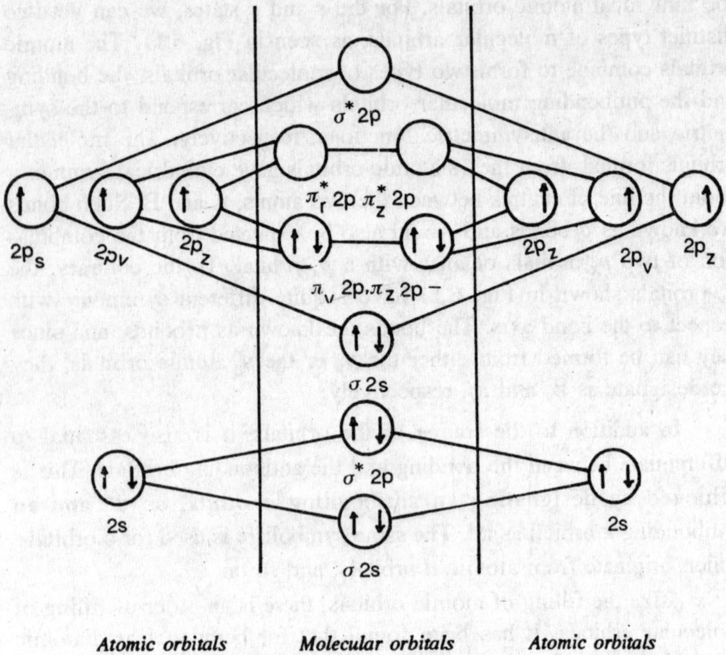

Fig. 6.12. *Molecular orbitals and their occupancy for the nitrogen molecule.*

In case of He_2, there are 4 electrons and it can be represented as

$$He\,[1s^2] + He[1s^2] \rightarrow He_2\,[(1_{1s})^2\,(\sigma^*{}_{1g})^2]$$

Here, we find that there are two electrons in the bonding orbital and two electrons in the antibonding orbital. The latter electrons cancel the effect of the bonding electrons, and so a stable He_2 molecule should not exist. However, the helium molecule ion, He_2^+, has been detected. Its stability can be attributed to an excess of bonding over antibonding electrons. An antibonding electron more than compensates for the bonding effect of a single bonding electron. Therefore, the helium molecule ion would not be expected to be 2*s* stable as a molecule having a simple one-electron bond.

Directional Bonding

Another and more outstanding accomplishments of the quantum mechnical approach to molecular structure is its success in the treatment of molecular geometry. Consider the bonds between the atoms in a molecule to be associated with the overlap of atomic orbitals, it is then expected that a definite geometry will be associated with the molecules. There are a number of ways to handle molecular structure. However, the use of localized atomic orbitals in terms of the valence bond theory has proved so successful that it is worthy of discussion.

In terms of the valence bond approach to molecular, structure, we can imagine a covalent bond to be formed because of the pairing of two electrons in the atomic orbitals of two different atoms. The bond, then should lie along the directions of the overlapping atomic orbitals and it is expected that the strongest bond will be formed in the position that allowed the maximum possible extent of overlap between the two orbitals. Let us consider the water molecules. The two hydrogen atoms use 1*s* orbital in the bond formation and the p_x and p_y orbitals of the oxygen atom. In Fig. 6.13 the *xy* plane of the water molecule is shown, and we find that the p_x and p_y orbitals are mutually perpendicular to each other in this plane. To obtain a maximum

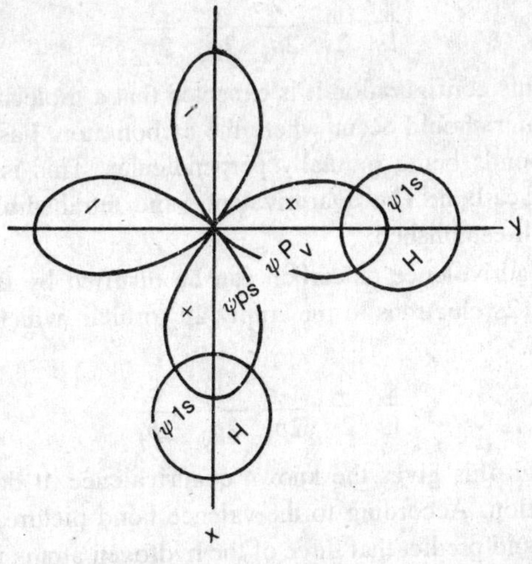

Fig. 6.13. *Overlap or Bonding Orbitals in the Water Molecule.*

possible overlap between the 1*s* orbitals of the hydrogen atoms and the p_x and p_y orbitals of the oxygen atom, it is necessary for the hydrogen atoms to approach along the *x* and *y* axes. This leads to the overlap shown, and the HOH bond angle should be 90°. This is not exactly in agreement with experiment. The angle has been found to be 104°31′. However, the deviation from 90° can be attributed to a repulsion between the two hydrogen atom and to the effect of some contribution to the bond of the 2*s* electrons of the oxygen atom. Thus the bond may also be treated as an sp^2 hybrid bond.

The same structure would be expected for H_2S, H_2Se, and H_2Te as is predicted for H_2O and in fact the agreement between experiment and theory is good for these molecules. The HXH angle for H_2S is observed to be 92.2°, for H_2Se it is 91.0°, and for H_2Te it is 89.5°. The same general argument may also be applied to the series NH_3, PH_3, AsH_3 and SbH_2. Here, all three of the *p* orbitals are involved in the bonding and the HXH bond angles should again be 90°. Experimentally, the angles are found for the series to be 107.3°, 93.3°, 91.8°, and 91.3°, respectively.

Hybridization

The simple orbital overlap approach work well for a variety of molecules, it fails completely for the organic compounds of carbon. The occupancy of the atomic orbitals of the carbon atom is

$$\underset{1s}{\uparrow\downarrow}\quad \underset{2s}{\uparrow\downarrow}\quad \underset{2p_x}{\downarrow}\quad \underset{2p_y}{\downarrow}\quad \underset{2p_z}{\downarrow}$$

From this configuration it is expected that a molecular formation similar to water should occur where the carbon atom has a valence of 2 with the bonds being mutually perpendicular. This is not, what is observed for carbon. The quardrivalence and tetrahedral structure of carbon is well-established.

The quadrivalence of carbon can be justified by the promotion of one of the 2*s* electrons to the empty $2p_z$ orbital, which leads to the configuration.

$$\underset{1s}{\uparrow\downarrow}\quad \underset{2s}{\downarrow}\quad \underset{2p_x}{\downarrow}\quad \underset{2p_y}{\downarrow}\quad \underset{2p_z}{\downarrow}$$

Although this gives the known quadrivalence, it does not agree with observation. According to the valence bond picture, such a configuration would predict that three of the hydrogen atoms in a molecule such as CH_4 are equivalent in energy and are also mutually perpendicular, but the fourth hydrogen atom is held by a weaker bond at an angle of about 125° with respect to the other bonds. But, this is contrary to the known geometry of the organic carbon atom.

However, it is possible to resolve the problem within the framework of the valence bond approach. We have seen that the wave function can be improved by taking linear combinations of several reasonable functions which described different representations or resonance structures of the system. Similarly, we might imagine that the actual wavefunctions for the four carbon bonds are represented by a mixture of the *s* and *p* orbitals available for bond formation. This character can be introduced into the various total wavefunctions by taking linear combinations of the *s* and *p* wavefuncitons. The best combination should be the one that would lead to the strongest bond.

Although it is quite reasonable to assume that the best wave function will be the one that gives the strongest bond, it is not obvious what basis we can use to measure the strength of the bond. It would therefore be reasonable, to assume that the strongest bond would be the one that allow the maximum possible overlap between the bonding orbitals. This is known as *the criterion of maximum overlap,* and is fundamental to the valence bond treatment of directional bonding.

sp$_3$ hybridization. In the formation of the hybrid orbitals, the radial portions of the orbitals in a given shell are approximately the same. This is apparent from a comparison of the radial portions of the 2*s* and the 2*p* orbitals. Thus, it is assumed that the hybridized orbitals can be formed from the angular parts of the individual wavefunctions. Thus, for the quadrivalent carbon atom, these should be four bonds of the form.

$$\Psi_1 = a_i\Psi_s + b_i\Psi_{px} + c_i\Psi_{py} + d_i\Psi_{pz} \qquad ...(6.43)$$

where the ψ_i are now purely angular wave functions.

Restricting ourselves to normalized functions, we can say that

$$\int \psi_i\psi_i^* \, d\tau = 1$$

Now, using the *s* orbital the charge can be imagined to be spread over a spherical distribution pattern. Introducding the element of volume for a sphere then gives

$$\iiint \psi_i\psi_i^* \, r^2 \sin\theta \, d\theta \, d\varphi \, dr = 1$$

Since we are assuming the radial portion of the wavefunction to be constant for all orbitals in the valence shell, our concern is with the angular portion of the wavefunction. Thus, we find that integration over the angular portion of the wavefunction results in

$$\int_0^{2\pi}\int_0^{\pi} \sin\theta \, d\theta \, d\varphi = 4\pi$$

Hence we say that the angualr portion of the wave function is normalized to 4π.

We shall neglect the radial portions of the wave functions so the individual angular wave functions can be expressed in terms of spherical harmonics so $\Psi_s = 1/\sqrt{4\pi}$ and the p orbitals all contain the term $\sqrt{3}/4\pi$. The result of the normalization is effectively to set Ψ equal to unity. This results in the following equations for the individual angular wavefunctions

$$\psi = 1$$
$$\psi_{px} = \sqrt{3}\sin\theta\cos\varphi$$
$$\psi_{py} = \sqrt{3}\sin\theta\sin\varphi$$
$$\psi_{pz} = \sqrt{3}\cos\theta$$

Due to the orthogonality of the various atomic orbitals, there is a restriction of the coefficients in Eq.(6.43)

$$a_i^{\,2} + b_i^{\,2} + c_i^{\,2} + d_i^{\,2} = 1$$

For determining the angular relationship between the bonds, we assume that the first bond lie along one of the axes, say the x axis. Then the p_y and p_z orbitals should make no contribution of the bond. So, c_i and d_i will both be zero and

$$\psi_1 = a_1\psi_s + b_1\psi_{pz}$$

Since

$$a_1^2 + b_1^2 = 1$$

Hence, we can say that

$$\psi_1 = a_1\psi_s + \sqrt{1 - a_1^2\ \psi_{px}}$$

Substituting the wavefunctions expressed in terms of ψ_s equal to unity, we get

$$\psi_1 = a_1 + \sqrt{1 - a_1^2}\,\sqrt{3}\sin\theta\cos\varphi$$

The bond will be formed in such a way that the strongest bond possible is formed and using the principle of maximum overlap, we can choose a_1 so as to make Ψ_1 as large as possible in the direction of the bond. Now, the value of Ψ_1 along the axis is given by

$$\psi_1 = a_1 + \sqrt{3}\,\sqrt{1 - a_1^{\,2}}$$

The $\sin\theta\cos\phi$ term is unity here because $\theta = \pi/2$ and $\varphi=0$ which are the maximum possible values for this term. Differentiating with respect to a_1 and putting the result equal to zero, we get

$$1 - a_1\sqrt{3}(1 - a_1^{\,2})^{-1/2} = 0$$

It can be seen that $a_1 = \frac{1}{2}$ and $b_1 = \sqrt{3/2}$, which gives us the wavefunction for the first hybrid bond, as

$$\psi_1 = \frac{1}{2} + (\sqrt{3/2}) \sin\theta \cos\varphi$$

which, in terms of the orbital symbols, can be expressed as

$$\psi_1 = \psi \frac{1}{2} + (\sqrt{3/2})\, \psi_{px}$$

If we evaluate their maximum magnitude in space on this basis by setting sin θ and cos φ each equal to unity, we get a value of 1.732 which indicates that a *p* orbital is capable of a greater amount of overlap than a pure *s* orbital, and according to the principle of maximum overlap, it should form a stronger bond. On the same basis we can find out the magnitude of the hybrid bond ψ_1. By putting sin θ and cos φ both equal to unity, a value of 2 is obtained for Ψ_1 in the direction of the bond. This is considerably greater than either a pure *s* or a pure *p* orbital, and so we expect the hybrid orbital to form a stronger bond than either a pure *s* or a pure *p* orbital.

The formation of the hybrid bond can be pictorially shown as in Fig. 6.14. The *s* orbital is everywhere positive and the *p* orbital has a positive and a negative lobe. The positive lobe of the *p* orbital combines with the *s* orbital and the negative lobe of the *p* orbital is reduced by the mixing.

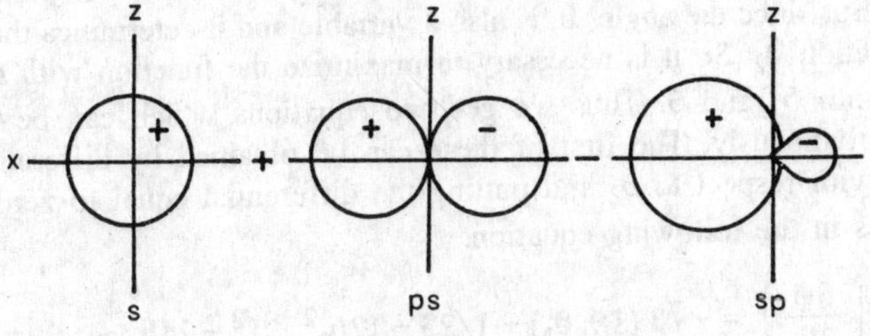

Fig. 6.14.

The second of the hybrid bonds can be chosen to lie in the *xz* plane. Here, the p_y orbital will make no contribution, and the new wavefunction will be given by

$$\psi_z = a_2\psi_8 + b_2\psi_{px} + d_2\psi_{py} \qquad \text{...(6.44)}$$

Since the orbital will lie in the *xz* plane so the angle φ will be

0° or 180° for the bond direction. However, for $\varphi = 0^{o}$, we are in the region of ψ_1. Therefore, Ψ_2 will have its maximum value when φ is 180°, and cos φ will be −1. Using this fact and substituting the expressions for the wavefunctions into Eq. (6.44), we get

$$\psi_2 = a_2 - b_2\sqrt{3}\sin\theta + d_2\sqrt{3}\cos\theta$$

This equation can be expressed in terms of the parameter b_2 and the angle θ by using the two relations

$$a_2^2 + b_2^2 + d_2^2 = 1$$

and

$$a_1 a_2 + n_1 b_2 = 0 \qquad ...(6.45)$$

The latter equation comes from the orthogonality condition. As d_l is zero so the d parameters vanish. If we know the values for a_1 and b_1 we can find out a_2 and d_2 in terms of b_2. Thus

$$a_2 = -(b_1/a_1)b_2 = -\frac{\sqrt{3/2}}{1/2}b_2 = -\sqrt{3}\,b_2$$

and

$$d_2 = \pm\sqrt{1 - a_2{}^2 - b_2{}^2} = \pm\sqrt{1 - 3b_2{}^2 - b_2{}^2} = \pm\sqrt{1 - 4b_2{}^2}$$

The sign chosen for $|\,d_2\,|$ ts immaterial but for convenience we use the positive sign. Thus we can express ψ_2 as

$$\psi_2 = -b_2\sqrt{3}\,(1 + \sin\theta) + \sqrt{3(1 - 4b_2{}^2)}\cos\theta$$

As in case of the first hybrid orbital, it is again necessary to maximize the orbital in the bond direction so as to find the parameter, b_2. But since the angle, θ, is also a variable and it determines the bond direction θ_2. So it is necessary to maximize the function with respect to both b_2 and θ. Thus we get two equations which can be solved simultaneously. The first of these can be obtained by differentiating ψ_2 with respect to b_2 and putting the differential equal to zero. This gives in the following equation.

$$\left(\frac{\partial\psi}{\partial b_2}\right)_\theta = -\sqrt{3}\,(1 + \theta_2) + 1/23 - 12b_2{}^{2-1/2} - 24b_3\cos\theta_2 = 0$$

or

$$-\sqrt{3(1 + \sin\theta_2)} - \frac{12b_2}{\sqrt{3 - 12b_2}}\cos\theta_2 = 0$$

Similarly,

$$\left(\frac{\partial\psi}{\partial\theta}\right)_{b2} = -b_2\sqrt{3}\cos\theta_2 - \sqrt{3 - 12b_2{}^2}\sin\theta_2 = 0$$

or $$- b_2\ 3\ \cos\theta^2 - \quad 3 - 12b_2{}^2 \sin\theta_2 = 0$$

These two equations can be used to find the angle, θ_2 and the parameter, b_2 arranging them we get

$$= \frac{12b_2}{\sqrt{3 - 12b_2{}^2}} = \frac{\sqrt{(31 + \sin\theta_2)}}{\cos\theta_2}$$

and $$= \frac{\sqrt{3}b_2}{\sqrt{3 - 12b_2{}^2}} = \frac{\sin\theta_2}{\cos\theta_2}$$

Thus

$$\frac{12\sin\theta_2}{\sqrt{3\cos\theta_2}} = \frac{\sqrt{3\,(1 + \sin\theta_2)}}{\cos\theta_2}$$

This equation can be solved for $\sin\theta_2$ to give $\sin\theta_2 = \frac{1}{3}$. Once we know $\sin\theta_2$ we can find b_2, and $b_2 = -\frac{1}{2}\sqrt{3}$. If these two quantities are both known, we can find the magnitude of the new hybrid orbital and we can determine its angle with respect to the first hybrid orbital. The magnitude is

$$\psi_2 = -\left(-\frac{1}{2\sqrt{3}}\right)\sqrt{3}\left(1 + \frac{1}{3}\right)$$

$$+ \sqrt{3 - 12\left(\frac{1}{2\sqrt{3}}\right)^2}\left(\frac{2\sqrt{2}}{3}\right) = 2$$

This is the same value that was obtained for ψ_1. Hence, we can say that the two bonds are of equal strength.

While considering the angle θ_2, remember that for the $\sin\theta_2$ is equal $\frac{1}{3}$, θ_{max} will equal 19° 28′ or 160° 32′. The former corresponds to choosing the positive sign for d_2. Since it was necessary to consider the angle φ to be 180°, it is apparent that, with respect to the first bond, the angle θ_2 is 90° + 19° 28′, or 109° 28′.

By a similar procedure we can determine the angles and magnitudes of the remaining two bonds. The general forms of the wave functions of these orbitals along with those of the first two orbitals are as follows :

$$\psi_1 = \frac{1}{2}\psi_8 + \sqrt{3/2}\,\psi_{pz}$$

$$\psi_2 = \frac{1}{2}\psi_8 - \frac{1}{2\sqrt{3}}\psi_{pz} - \frac{\sqrt{3}}{\sqrt{3}}\psi_{p^z}$$

$$\psi_3 - \frac{1}{2}\psi_8 - \frac{1}{2\sqrt{3}}\psi_{pz} + \frac{1}{\sqrt{2}}\psi_{pz} - \frac{1}{\sqrt{6}}\psi_{pz}$$

$$\psi_4 = \frac{1}{2}\psi_8 - \frac{1}{2\sqrt{3}}\psi_{pz} - \frac{1}{\sqrt{3}}\psi_{pz} + \frac{1}{\sqrt{6}}\psi_{pz}$$

Thus the angles all turn out to be 109° 28′, and the relative magnitude are all 2. In the methane molecule, we know that all four of the bonds are equivalent and the HCH angles are all 109° 28′. Thus, it would appear that the valence bond theory can give a satisfactory explanation of the valence bond theory can give a satisfactory explanation of the tetrahedral carbon atom. The tetrahedral hybrid bonds are formed from one *s* orbital and three *p* orbitals, and so they are known as *sp^3 hybrids.*

Sp^2 and sp hybridization. The treatment of double-bonded structure is in terms of sp^2 hybrid orbitals. For a molecule like ethylene,

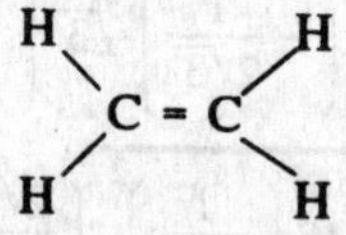

we can imagine the *skeletal* structure to be

From this we find that there are three sigma bonds from each carbon atom. These can be considered to arise from the 2*s* and two of the three 2*p* electrons of the carbon atom in the form of an sp^2 hybridization, leaving one unused 2*p* electron on each carbon atom. By hybridizing these three orbitals, three equivalent bonds are formed which lie in a plane at an angle of 120° with respect to one another. This is in agreement with experiment. Their magnitude compared to unity for an *s* orbital is 1991 which is less than that of an sp^2 hybrid, but it is considerably greater than that of a pure *p* orbital or a pure *s* orbital.

The double-bond character of the carbon-carbon bond can be attributed to the overlap of the *p* orbitals occupied by the unused *p* electrons. This is illustrated in Fig. 6.15. The overlap of the sp^2 hybrid orbitals leads to the sigma bonds and the overlap of the *p* orbitals leads to a pi bond. From the figure 6.15 it is clear why restricted rotation should exist around a double bond. For one of the carbon atoms to rotate around an axis through the line of centres of the two carbon atoms, it is essential to break the overlap of the π orbitals.

For the acetylenic-type linkage, the skeletal structure could be

$$H - C \equiv C - H$$

This structure requires the use of only one of the $2p$ orbitals along with the $2s$ orbital to form the hybrid bond. Thus, we have two p electrons on each carbon atom which are unused in the formation of the sigma bond. The calculations of the geometry and relative strengths of two equivalent sp hybrids are again found to be in agreement with the experimentally determined structure of the molecule. Both bonds are of equal strength and they are linear. The triple-bond character of

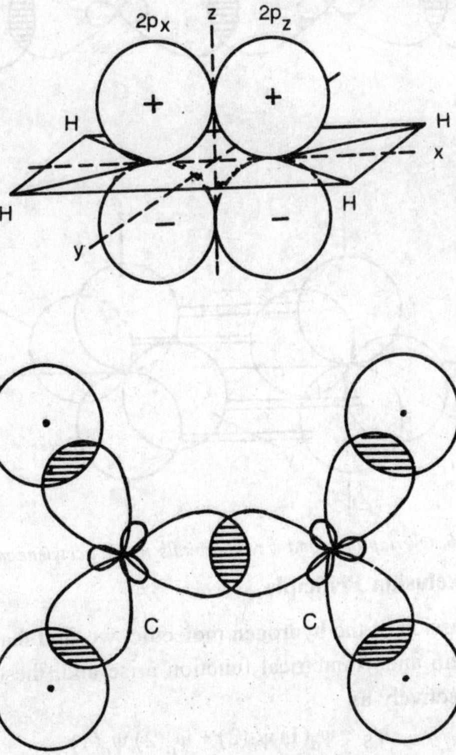

Fig. 6.15. *Overlap of σ and π-Type orbitals in the ethylene molecule.*

the acetylenic linkage can then the attributed to the overlap of the two *p* orbitals that are left on each carbon atom, as can be seen in Fig. 6.16.

Hybridization with d Orbitals

The valence bond theory has also proved moderately successful in its treatment of the structure of coordination compounds. For this it is necessary to consider the effects of the *d* orbitals. The most common structures which result from the hybridization of *d* orbitals are the dsp^2 hybrid bonds, which lead to square planar structures, and the d^2sp^2 hybrid bonds which lead to octahedral structures.

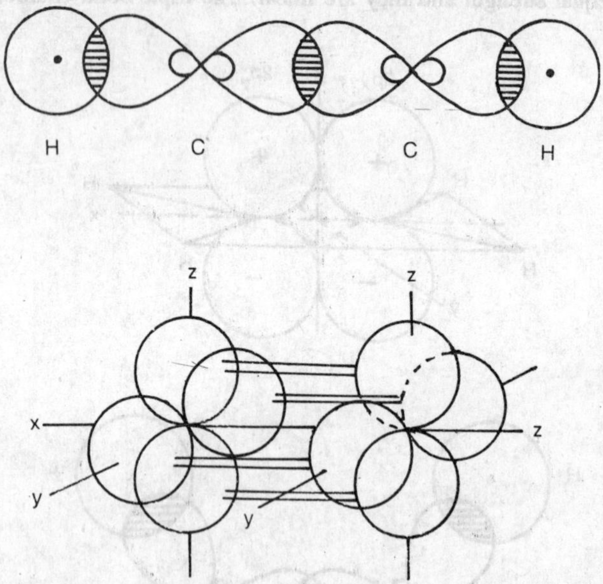

Fig. 6.16. *Overlap of σ and π type orbitals in the acetylene molecule.*

The Pauli Exclusion Principle

In discussion of the hydrogen molecule, we find that both a symmetrical and an antisymmetrical function arise and, these can be represented respectively as

$$\psi_S = \psi_a(1)\,\psi_b(2) + \psi_a(2)\,\psi_a(1)$$

and

$$\psi_A = \psi_a(1)\,\psi_b(2) - \psi_a(2)\,\psi(1)$$

If we take the electrons in the system described by these two wave functions and exchange the coordinates of electron (1) with the coordinates of electron (2), we will get for the symmetric function

$$\psi_S' = \psi_a(2)\,\psi_b(1) + \psi_a(1)\,\psi_b(2)$$

But, this is the same as $\Psi_{S.}$ Hence on exchange of coordinates, we find that $\psi_S = \psi_S'$. Now if we exchange the coordinates of the electrons in the antisymmetrical function, we get

$$\psi_A' = \psi_a(2)\,\psi_b(1) - \psi_a(1)\,\psi_b(2)$$

This is not the same but is opposite in sign from ψ_A. Therefore in this case, we find that for the exchanged of coordinates, $\psi_A = -\psi_A''$.

So far we have considered only the *orbital wavefunctions,* and these are based on the spatial quantum numbers *n, l* and *m,* so we have neglected the *spin wavefunctions.* For a simple discussion of valency, it is possible to separate the spin wavefunctions from the orbital wave functions because spin-orbit interactions are generally negligible. However, the electron spin must be included in any complete description of the system.

As an electron can a spin of either $+\frac{1}{2}$ or $-\frac{1}{2}$, which can be represented as α and β, repectively. If both electrons in an electron pair have spins of $+\frac{1}{2}$, then their combined wave function will be $\alpha(1)\,\alpha(2)$. If both of the electrons have a spin of $-\frac{1}{2}$, then their combined wave functiin will be $\beta(1)\,\beta(2)$. Finally, if one of the electrons has a spin of $+\frac{1}{2}$ and the other has a spin of $-\frac{1}{2}$, the combined wave function will be

$$\alpha(1)\,\beta(2) \pm \alpha(2)\,\beta(1)$$

The liner combination of wavefunctions arises here due to the indistinguishability of the electrons. Further, if we exchange the electrons we find that the function

$$\alpha(1)\,\beta(2) + \alpha(2)\beta(1)$$

is a symmetical function, but the function

$$\alpha(1)\,\beta(2) - \alpha(2)\,\beta(1)$$

is an antisymmetricals function.

Now if we consider both the spatial portion and the spin portion of the system, it is essential to combine these to give the total wavefunction including spin. As was seen earlier the total wave function for the hydrogen atom can be expressed as a product of the individual wavefunctions, such that

$$\psi_r = R_{(r)}\,\theta_{(\theta)}\Phi_{(\varphi)}$$

Now if we introduce a fourth degree of freedom, the product relationship must still hold. Thus, the total wavefunction for the hydrogen molecule, including spin, should be formed from the product of the orbital function and the spin function. That is, for this system

$$\psi_T = \{\psi_a(1)\,\psi_b(2) \pm \psi_a(2)\,\psi_b(1)\}\ \text{(spin functions)}$$

We can see that this leads to the eight possible combinations

$$\{\psi_a(1)\,\psi_b(2) + \psi_a(2)\,\psi_b(1)\}\begin{Bmatrix} \alpha(1)\,\beta(2) - \alpha(2)\,\beta(1) \\ \alpha(1)\,\alpha(2) \\ \beta(1)\,\beta(2) \\ \alpha(1)\,\beta(2) + \alpha(2)\,\beta(1) \end{Bmatrix}$$

$$\{\psi_a(1)\,\psi_b(2) + \psi_a(2)\,\psi_b(1)\begin{Bmatrix} \alpha(1)\,\alpha(2) \\ \beta(1)\,\alpha(2) \\ \alpha(1)\,\beta(2) + \alpha(2)\,\beta(1) \\ \beta(1)\,\beta(2) - (2)\alpha\,\beta(1) \end{Bmatrix}$$

Now our aim is to find whether all eight of these combinations are significant. The Pauli exclusion principle first enunciated in 1925, which states, that *no two electrons in the some system can have the same values for all four quantum numbers provides the answer.* In terms of quantum mechanics, however, it takes on the more subtle forin and may be stated as *every total wavefunction must be antisymmetric in the exchange of the coordinates of every pair of electrons.* In terms of this restriction, only four of the eight combinations will be acceptable. These are

$$\{\psi_a(1)\,\psi_b(2) + \psi_a(2)\,\psi_b(1)\}\,\{\alpha(1)\,\beta(2) - \alpha(2)\,\beta(1)\}$$

$$\{\psi_a(1)\,\psi_b(2) - \psi_a(2)\,\psi_b(1)\}\begin{Bmatrix} \alpha(1)\,\alpha(2) \\ \beta(1)\,\beta(2) \\ \alpha(1)\,\beta(2) + \alpha(2)\,\beta(1) \end{Bmatrix}$$

Hence, it appears that there are three states corresponding to the anti-symmetric spatial function, where there is only one state corresponding to the symmetric spatial function. We refer to the ground state as a *singlet state* and to the repulsive or antisymmetric state as a *triplet state.*

The relationship between the quantum mechanical statement of the exclusion principle and the more familiar statement in terms of the four quantum numbers may not be evident. But, it can be shown that the original statement of the exclusion principle is a necessary feature of the more general statement of the principle in terms of symmetry. If we take thc total ground-state function

$$\phi_T = \{\psi_a(1)\,\psi_b(2) + \psi_a(2)\,\psi_a(1)\}\,\{\alpha(1)\,\beta(2) - \alpha(2)\,\beta(1)\}$$

we get on multiplication

$$\psi_T = \psi_a(1)\alpha(1)\,\psi_b(2)\beta(2) - \psi_a(2)\alpha(2)\,\psi_b(1)\beta(1) + \psi_a(2)\,\beta(2)\,\psi_b(1)\alpha(1) - \psi_a(1)\beta(1)\psi_b(2)\alpha(2)$$

It can be expressed more conveniently in the determinantal form as

$$\psi_T = \begin{vmatrix} \psi_a(1)\alpha(1) & \psi_b(1)\psi(1) \\ \psi_a(2)\alpha(2) & \psi_b(2)\beta(2) \end{vmatrix} - \begin{vmatrix} \psi_a(1)\beta(1) & \psi_b(1)\alpha(1) \\ \psi_a(2)\beta(2) & \psi_b(2)\alpha(2) \end{vmatrix}$$

Now if we consider the three quantum numbers *n, l* and *m* to be identical for both elctrons, then $\Psi_\alpha = \Psi_{\beta,}$ and the determinant can be expressed as

$$\psi_T = \begin{vmatrix} \psi_a(1)\alpha(1) & \psi_b(1)\beta(1) \\ \psi_a(2)\alpha(2) & \psi_b(2)\beta(2) \end{vmatrix} - \begin{vmatrix} \psi_a(1)\beta(1) & \psi_b(1)\alpha(1) \\ \psi_a(2)\beta(2) & \psi_b(2)\alpha(2) \end{vmatrix}$$

or

$$\psi_T = 2\begin{vmatrix} \psi_a(1)\alpha(1) & \psi_b(1)\beta(1) \\ \psi_a(2)\alpha_{(2)} & \psi_b(2)\beta(2) \end{vmatrix}$$

Now, if we allow α to equal β, the two rows of the determinant become identical and the determinant vanishes. Thus, the wave function is seen to be equal to zero when all four quantum numbers are the same for two different electrons in the same system. And this, of course, is what is demanded by the Pauli exclusion principle.

7

Hückel Molecular Orbital Theory

HÜCKEL DIAGRAMS

We already have enough mathematical appratus to construct a poor man's quantum mechanics for planar, aromatic hydrocarbons. While the methods employed bear only a superficial resembblance to the more sophisticated quantum mechanics that we develop later, the results are nevertheless capable of reproducing approximately many of the properties of these compounds which are familiar to the organic chemist.

We start with a model (Hückel diagram) of the hydrocarbon consisting of carbon atoms (circles) and bonds (lines) drawn in such a way as to reproduce the molecular structure formulas of the organic chemist. Some samples are shown in Figure 7.1, and we shall later learn that the bonds are to be identified with the π bonds of a conjugated, organic molecule.

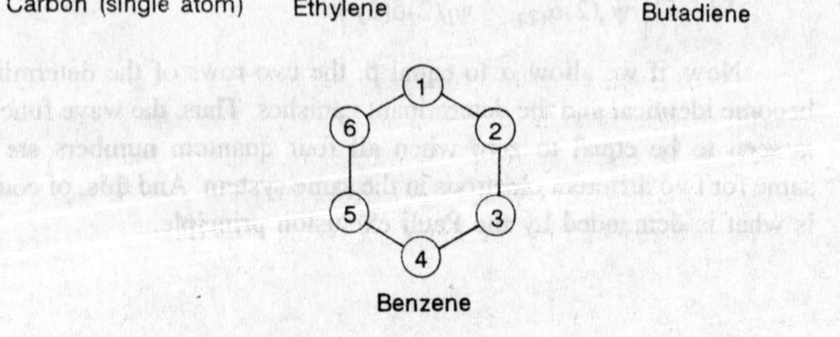

Fig. 7.1. *Some Hückel diagrams.*

Simplified Quantum Postulates

The atoms in the Hückel diagram are numbered in any arbitrary way. Into the atomic framework we feed electrons, and we postulate that the electrons are distributed over the atoms according to a probability law which may be calculated as follows:

1. The "state" of an electron shall be said to be determined by a *molecular orbital* defined to be a vector $\Psi = (c_1, c_2, \ldots.., c_n)$ in an n dimensional vector space where n is the number of carbon atoms in the compound. The components c_j of the vector have the interpretation that $c_i{}^2$ is the probability of observing the electron on the jth atom. Because the probability of observing the electrons somewhere in the molecules is unity, we must have $c_1{}^2 + c_2{}^2 + \ldots.. + c_n{}^2 = 1$, or, in vector notation,

$$\psi^T\psi = 1$$

2. The acceptable vectors that may be used to describe the state of an electron are the eigenvectors of an n dimensional matrix H called the Hamiltonian matrix. This matrix is to be constructed in such a way that $H_{ii} = \alpha$; $H_{ij} = \beta$ if $i \neq j$ and atoms i and j are bonded; $H_{ij} = 0$ otherwise. The parameters α and β have the dimensions of energy, so that eigenvalues E associated with eigenvectors ψ also have the dimensions of energy, and an eigenvalue is interpreted to be the energy of an electron in the state ψ. Note that H is symmetric $H_{ij} = H_{ij}$, so that the ψ_i will form an orthonormal set of vectors.

3. The exclusion principle: if a molecule possesses more than one electron distributed over its atoms, the same state vector may be used to describe the state of not more than two electrons at a time. Other electrons must then be described by other state vectors. Finally, the total electron density at atom j is obtained by summing at that atom the contributions $c_j{}^2$ from all electrons in the molecule.

The Carbon Atom

To get an idea of how these rules work, let us examine the simplest possible case, that of a single carbon atom (Fig. 7.1). Because $n = 1$, our vector space is one dimensional, $\psi = c_1$, and H consists of the single element, $H_{11} = \alpha$. The eigenvectors of H satisfy

$$H\psi = E\psi$$

or

$$\alpha c_1 = E c_1$$

whence $E = \alpha$ is the single energy available to an electron in a carbon atom. The corresponding eigenvector after normalization is

$$\psi = 1$$

so that $c_1 = 1$, and the probability of observing the electron on the atom is $1^2 = 1$. All of this is trivially simple, but it does yield a useful interpretation of α, telling us that α is the energy of an alectron in an isolated carbon atom It is customary to take the zero of energy for quantum mechanical systems as that when all the particles involved are at infinite separation. With this convention α must be negative, for the carbon atom is electronically stable, and its bond electron must lie at an energy level lower than zero.

Ethylene

Referring to Figure 7.1, our vector space is of dimension $n = 2$, and we set up matrix elements

$$H_{11} = H_{22} = \alpha; \quad H_{12} = H_{21} = \beta$$

Then

$$H\psi = E\psi$$

or

$$\begin{pmatrix} \alpha & \beta \\ \beta & \alpha \end{pmatrix} \begin{pmatrix} c_1 \\ c_2 \end{pmatrix} = E \begin{pmatrix} c_1 \\ c_2 \end{pmatrix}$$

The secular equation is

$$\begin{vmatrix} \alpha - E & \beta \\ \beta & \alpha - E \end{vmatrix} = 0$$

whence we have eigenvalues $E_1 = \alpha + \beta$; $E_2 = \alpha - \beta$. To obtain results corresponding to experiment, it turns out that we must empirically assume that β like α is also negative, so that the most stable energy level for an electron in ethylene is E_1, and the neutral molecule may accommodate both of its electrons in this level without violating the exclusion principle. Thus the total ground state energy for the neutral molecule is $2E_1 = 2\alpha + 2\beta$.

To compare this molecular energy with the atomic energy of two unbonded carbon atoms, rub out the bond between 1 and 2 and calculate the energy for the Hückel diagram;

① ②

The modified Hamiltonian matrix becomes

$$H = \begin{pmatrix} \alpha & 0 \\ 0 & \alpha \end{pmatrix}$$

whence the unbonded atoms possess a doubly degenerate energy level $E_1 = E_2 = \alpha$, so that two electrons in this level have a total energy 2α. Comparing this with the molecular energy of ethylene, it is evident that the bond energy (read π bond energy) in ethylene is 2β, which gives us a physical interpretation of β as one half the energy of an olefinic π bond.

Figure 7.2 illustrates the change in energy levels upon molecule formation in ethylene. The molecular energy levels are classified as bonding or antibonding according to whether they lie below or above the original atomic levels before molecule formation

Associated with E_1 in ethylene is an eigenvector $\psi_1 = (1/\sqrt{2},$

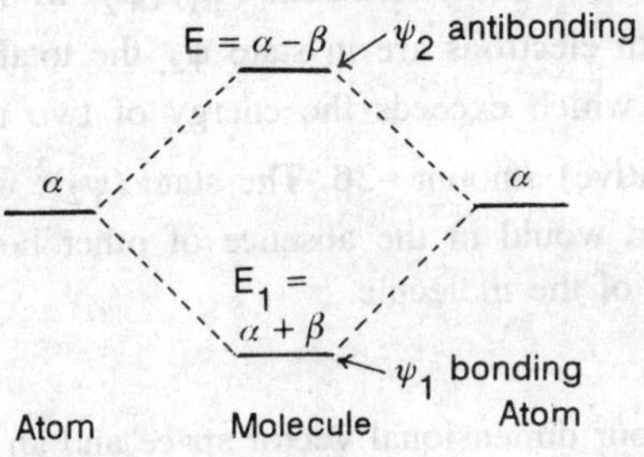

Fig. 7.2. Comparison of atomic and molecular levels in ethylene.

$1/\sqrt{2})$ and with E_2 an eigenvector $\psi_2 = (1/\sqrt{2}, -1/\sqrt{2})$. Because both electrons in the ground state of the molecule are described by ψ_1, each electron makes a contribution of $(1/\sqrt{2})^2 = \frac{1}{2}$ to the total electron density at each carbon, whence there is a total of $\frac{1}{2} + \frac{1}{2} = 1$ electron density at each carbon, atom. The electronic configuration of two electrons in ψ_1 is conventionally denoted by the symbol $(\psi_1)^2$.

Note that the same electron density at each carbon would be obtained from ψ_2, and at first sight this would appear to be inconsistent with the fact that ψ_1 possesses a lower energy, If, however, the electrons were conceived to be continuously distributed throughout the special region in the neighborhood of the atoms, then the change in sign between c_1 and c_2 in ψ_2 would imply a region between the atoms of zero electron density—a node as we shall later learn to call it. ψ_1 exhibits no such change in sign, and we may infer a nonvanishing electron density between the atoms. These conclusions are in accord with the chemist's traditional picture of the formation of a chemical bond as being accompanied by the buildup of electron density between the bonded atoms and constitute our empirical reason for taking β to be negative.

If one electron is in ψ_1 and the other is in ψ_2, we speak of the molecule as being in an excited state and write its configuration as $(\psi_1)(\psi_2)$, but according to our present formalism, this excited state would have a total energy $E_1 + E_2 = 2\alpha$, which is the same as two isolated carbon atoms. The excited state $(\psi_1)(\psi_2)$ would therefore be nonbonding. If both electrons are in state ψ_2, the total energy would be $2E_2 = 2\alpha - 2\beta$, which exceeds the energy of two isolated carbon atoms by the (positive) amount -2β. The state $(\psi_2)^2$ would therefore be antibonding and would in the absence of other bonding for sces lead to dissolution of the n.olecule.

Butadiene

We have a four dimensional vector space and an H matrix with $H_{11} = H_{22} = H_{33} = H_{44} = \alpha$; $H_{12} = H_{23} = H_{34} = \beta$; $H_{13} = H_{14} = H_{24} = 0$. The secular equation is

$$\begin{vmatrix} \alpha - E & \beta & 0 & 0 \\ \beta & \alpha - E & \beta & 0 \\ 0 & \beta & \alpha - E & \beta \\ 0 & 0 & \beta & \alpha - E \end{vmatrix} = 0 \quad \text{...(7.1)}$$

The roots of this equation are unaffected if β is factored out of each row, and it is also convenient to define $x = (\alpha - E)/\beta$. Then

$$\begin{vmatrix} x & 1 & 0 & 0 \\ 1 & x & 1 & 0 \\ 0 & 1 & x & 1 \\ 0 & 0 & 1 & x \end{vmatrix} = 0$$

Expansion of the determinant leads to a secular polynomial

$$x^4 - 3x^2 + 1 = 0$$

with roots

$$x = \pm\left(\frac{3}{2} \pm \frac{1}{2}\sqrt{5}\right)^{1/2}$$

The allowed energy levels are therefore ($E = x - x\beta$):

$$\begin{aligned} E_1 &= \alpha + 1.618\beta \\ E_2 &= \alpha + 0.618\beta \\ E_3 &= \alpha - 0.618\beta \\ E_4 &= \alpha - 1.618\beta \end{aligned} \qquad ...(7.2)$$

The most stable configuration for the neutral molecule will arise will arise if we place two electrons in each of the two lowest energy levels E_1 and E_2, thus leading to a total of $E = 2E_1 + 2E_2 = 4\alpha + 4.472\beta$. It is instructive to note that not only is this energy lower by a total energy 4.472 β than the energy of four isolated carbons,

① ② ③ ④

but it is also lower by 0.472 β than the energy of two isolated ethylenes with Hückel diagram

①------------② ③------------④

and secular equation

$$\begin{vmatrix} \alpha - E & \beta & 0 & 0 \\ \beta & \alpha - E & \beta & 0 \\ 0 & 0 & \alpha - E & \beta \\ 0 & 0 & \beta & \alpha - E \end{vmatrix} = \begin{vmatrix} \alpha - E & \beta \\ \beta & \alpha - E \end{vmatrix}^2 = 0$$

This increased stabilization energy given to a conjugated molecule by delocalization of its double bonds is called its *delocalization energy*.

Eigenvectors corresponding to each of the eigenvalues (7.2) are

$$\begin{aligned} \psi_1 &= (0.372, \quad 0.601, \quad 0.601, \quad 0.372) \\ \psi_2 &= (0.601, \quad 0.372, \quad -0.372, \quad -0.601) \\ \psi_3 &= (0.601, \quad -0.372, \quad -0.372, \quad 0.601) \\ \psi_4 &= (0.372, \quad -0.601, \quad 0.601, \quad -0.372) \end{aligned} \qquad ...(7.3)$$

When the first two levels of butadiene are filled with two electrons each, we write the configuration $(\psi_1)^2(\psi_2)^2$, and there will be $2(0.372)^2 + 2(0.601)^2 = 1$ electron on the average on each carbon. There are no nodes or changes of sign in the components of ψ_1, which is characteristic of the most stable electronic states. The two electrons in ψ_2, however, each have a node between atoms 2 and 3, and from our experience with ethylene, this would suggest that some of the bonding between 2 and 3 which is provided by ψ_1 is cancelled out by ψ_2, which at the same time reinforces the bonding between atoms 1 and 2 and atoms 3 and 4. In other words, butadiene should have more double bond character for the outer two bonds than for the inner one. This conclusion can be made quantitative by defining what is known as the *bond order* P_{rs} for the two bonded atoms r and s. This is computed by forming the products $c_r c_s$ taken from every occupied state vector of the molecule for the fixed positions r and s and summing the contributions from each electron.

Thus for ethylene, both electrons in the ground state are in $\psi_1 = (1/\sqrt{2}, 1/\sqrt{2})$. Hence $c_1 c_2 = (1/\sqrt{2})(1/\sqrt{2}) = \frac{1}{2}$ is the contribution from each electron, so that the total bond order is $\frac{1}{2} + \frac{1}{2} = 1$, and we may take this as the bond order of a true, olefinic double bond. For butadiene in the ground state, ψ_1 and ψ_2 are each occupied by two electrons. The bond between atoms 1 and 2 has a contribution $c_1 c_2 =$ (0.372) (0.601) from ψ_1 and a contribution $c_1 c_2 =$ (0.601)(0.372) from ψ_2. The total bond order for this bond is thus 2(0.372) 0.601) + 2(0.601) (0.372) = 0.894, somewhat less than the ethylenic double bond.

Similarly, the bond order for the 2-3 bonds is $2c_2c_3$ from ψ_1 + $2c_2c_3$ from ψ_2, or 2(0.601)(0.601) + 2(0.372)(–0.372) = 0.447–an even weaker double bond.

Benzene

Our exprience by now is such that we may rapidly set up the secular determinant.

$$\begin{vmatrix} x & 1 & 0 & 0 & 0 & 1 \\ 1 & x & 1 & 0 & 0 & 0 \\ 0 & 1 & x & 1 & 0 & 0 \\ 0 & 0 & 1 & x & 1 & 0 \\ 0 & 0 & 0 & 1 & x & 1 \\ 1 & 0 & 0 & 0 & 1 & x \end{vmatrix} = 0$$

where, as before, $x = (\alpha - E)/\beta$. The roots prove to be $x = -2, -1, -1, 1, 1, 2$, and we note that for this molecule two of the energy levels are degenerate. In order of increasing energy, the eigenvalues are

$$E_1 = \alpha + 2\beta$$
$$E_2 = E_3 = \alpha + \beta$$
$$E_4 = E_5 = \alpha - \beta$$
$$E_6 = \alpha - 2\beta$$

and the eigenvectors prove to be

$$\begin{aligned}
&\psi_1 = (1/\sqrt{6})\,(1, 1, 1, 1, 1, 1)\\
&\left\{\begin{aligned}\psi_2 &= (1/\sqrt{4})\,(0, 1, 1, 0, -1, -1)\\ \psi_3 &= (1/\sqrt{12})\,(2, 1, -1, -2, -1, 1)\end{aligned}\right.\\
&\left\{\begin{aligned}\psi_4 &= (1/\sqrt{4})\,(1, -1, 0, 1, -1, 0)\\ \psi_5 &= (1/\sqrt{12})\,(1, 1, -2, 1, 1, -2)\end{aligned}\right.\\
&\psi_6 = (1/\sqrt{6})\,(1, -1, 1, -1, 1, -1)
\end{aligned} \quad \text{...(7.4)}$$

in which I have bracketed the degenerate sets. Because of the degeneracy, this is only one of many possible sets of eigenvectors, and any normalized linear combination of ψ_2 and ψ_3 is also an eigenvector, as in any normalized linear combination of ψ_4 and ψ_5. Note that I have chosen the eigenvectors spanning the two dimensional degnerate sub-spaces to be mutually orthogonal, so that the entire set of six vectors is an orthonormal basis in the vector space of six dimentions.

The neutral benzene molecule will have six electrons to be accommodated in these energy levels, and our exclusion principle requires that in the ground state, two electrons will reside in each of ψ_1, ψ_2, and ψ_3. The total energy of the configuration $(\psi_1)^2(\psi_2)^2(\psi_3)^2$ will thus be $E = 2E_1 + 2E_2 + 2E_3 = 6\alpha + 8\beta$. Three isolated double bonds (cyclohexatriene) would have an energy $6\alpha + 6\beta$, and the difference 2β between the two is the delocalization energy of benzene.

To obtain the electron density at each atom, sum the squares of corresponding vector components over all electrons. Thus the electron density at atom 1 is $2[(1/\sqrt{6})^2 + 0^2 + (2/\sqrt{12})^2] = 1$, and the same value will be found for each atom in the molecule.

There is no difficulty in interpreting ψ_1 as being a bonding state vector, for it has no nodes anywhere in the molecule. ψ_2 and ψ_3 have two nodal points each, however, and upon inspection these prove to lie on two mutually perpendicular nodal lines passing through the molecule (Fig. 7.3). Despite the antibonding character of these nodes, the energy $\alpha + \beta$ of an electron in either of ψ_2 and ψ_3 indicates that it adds overall stability to the molecule. It might at first appear that the location of these nodes would permit different atoms to have different properties. Should not, for example, atoms 1 and 4 which lie on a nodal line have properties different from atoms 2 and 3 or 5 and 6 which straddle a nodal line? The answer is no, for so long as they are mutually perpendicular, the nodal lines have an arbitrary orientation with respect to the molecule. Other orientations correspond to other pairs of mutually orthogonal vectors belonging to the eigenvalue $x + \beta$. Thus ψ_2 and ψ_3 may be replaced by

$$\psi_2' = (1/\sqrt{4})(1, 1, 0, -1, -1, 0)$$

$$\psi_3' = (1/\sqrt{12})(1, -1, -2, -1, 1, 2)$$

All of our postulates are still satisfied, but the nodal lines now have the appearance of Figure 7.4. The quantum mechanical interpretation of this situation is that while we are permitted to know of existence of a nodal line, the degeneracy of this energy level probibits us from identifying its orientation.

I emphasize this arbitrariness by pointing ψ_2' and ψ_3' above are only one out of infinitely many orthonormal pairs of vectors that can be assigned to the degenerate level $x + \beta$. In general, any pair of linear combinations

$$\psi_2 = a_{22}\psi_2 + a_{23}\psi_3$$

$$\psi_3 = a_{32}\psi_2 + a_{33}\psi_3$$

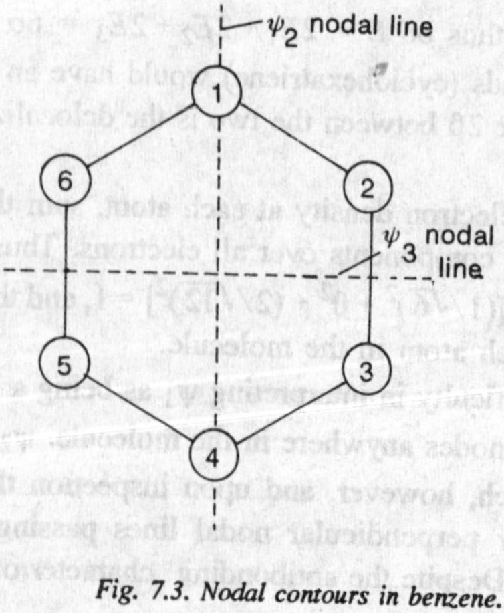

Fig. 7.3. Nodal contours in benzene.

for which the a_{ij} are the components of an orthogonal matrix are equally valid eigenvectors for this level, and each distinct such pair will rotate the nodal patterns of ψ_2 and ψ_3 through an arbitrary angle. This is a general phenomenon in quantum mechanics, and we shall see that alternative sets of degenerate eigenvectors can be transformed into each other by means of an orthogonal matrix.

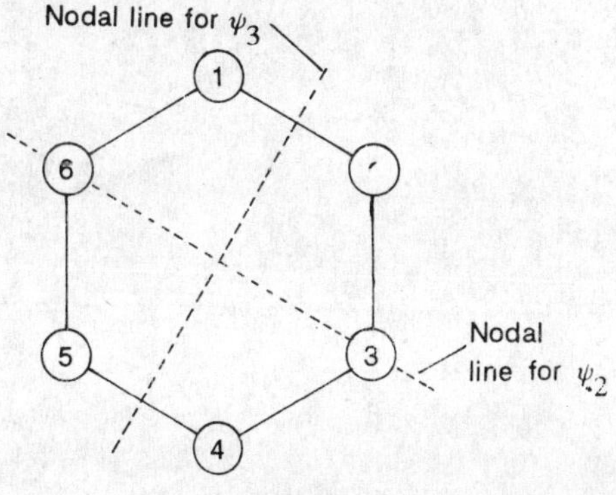

Fig. 7.4. *Alternative nodal contours in benzene.*

Also important to us as scientists is the realization that quantum mechanics frequently confronts the investigator with undecidable propositions, such as the orientation of the nodes of the degenerate $\alpha + \beta$ molecular orbitals of benzene. It is a measure of the maturity of the student's grasp of our subject that he acept the existence of such undecidable propositions along with the realization that should he insist on imposing a decision on nature *philosophically,* none of the *experimental* properties (energy, electron density distribution, etc.) of the quantum mechanical system will be changed.

The indecidable propositions are thus removed from the realm of experimental confirmation into that of philosophic speculation, a region not forbidden to the scientist, but into which he enters stripped of any particular authority.